Contents

^^

Arc welding

Arc welding is both a excellent trade to learn and master, and a art form in the right hands. This method of fusion between to pieces of metal into one, is truly a miracle of the modern industrialized world.

There are several methods of joining metals other than by means of mechanical fastenings, such as bolts, clamps, rivets, hinges, etc. The first form of jointure known to man, other than the above-mentioned, was fire or forge welding performed by a smith, the operation consisting essentially of heating the parts to be welded to the proper temperature and perfecting a union by applying pressure by means of hammer and anvil.

As pressure welding was limited in its application man endeavored to find some other way to join metals, or to make additions of metal to other metal, without the use of pressure. He was eventually successful in this endeavor and welding without pressure came to be known as autogenous welding, so called because of its self- or auto-generation ; i. e., it is self-produced by the application of intense heat without any physical process of compression or hammering.

Therefore, the two general forms of welding one requiring external application of pressure to complete the weld, and one in which the weld is completed without the external application of pressure.

In the first form the union is secured by using a comparatively low heat and high pressure. In the second the union is secured by a relatively high temperature without the aid of any external pressure.

It will be seen, in view of the fact that welding requires actual fusion of the metals joined or added, that the process differs inherently from those methods of joining metals known as brazing or soldering, in which cold surfaces are united by the interposition of a fused metallic cementing material, which is an example of adhesion rather than cohesion. **Pressure Welding.** *The conditions for successful smith or forge welding, which is a form of pressure welding, may be summed up as clean metallic surfaces in contact, with a suitable temperature and rapid closing of the joints. All the variations in the forms of welds are due either to differences in shapes of material or to the different practices of different craftsmen.*

The typical weld is the scarf; the joint is made diagonally to give a long contact at the point of union. Abutting faces are made slightly convex. The object is to allow any scale or dirt to be forced out which if allowed to become embedded in the joint would impair its union. It is important to have the proper temperature or else the metal will become badly oxidized (burnt) and will not adhere. This is especially true in the case of steel welding.

Resistance Welding *. Resistance welding is another form of pressure welding in which an electric current is made use of to produce the welding heat. There are two general forms of resistance welding; namely, butt and spot, the name in each case being thoroughly indicative*

of the service for which each form is particularly adapted.

Butt welding *is accomplished by having the surfaces, or parts of the metal to be united, fitted approximately to each other. Clamps of suitable design, generally made of copper, are then attached in as close proximity to the weld as is practicable and in such a way as to permit the desired amount of current to pass through the parts to be joined or welded. The resistance offered to the passage of the current at the point of contact produces the welding heat; whereupon sufficient pressure is applied to effect the union.*

Spot welding *also utilizes the heat generated by the resistance offered to the passage of an electric current and is similar to butt welding except that heat is generated at the points of contact between the respective electrodes in addition to the heat generated between the surfaces to be united.*

Seam welding *utilizes the heat generated in a way quite similar to spot welding; in fact, it is an extension of spot welding. A spot weld is equivalent in form to flush riveting; a seam weld is a non-interrupted continuous succession of spot welds.*

All forms of resistance welding require primarily a heavy current at a low potential, which practically necessitates the use of alternating current. The welding equipment, therefore, generally consists of a step-down transformer with a regulating device; clamps or electrodes for making the electrical connections to the work; and suitable mechanical parts and devices for supporting the electrodes, supplying pressure to the weld and supporting the parts to be welded.

In general, it may be said that the resistance form of welding is best adapted to standardized operations, especially so in the manufacturing field where the work can be passed through the machine.

While it might seem that the application of this form of welding is somewhat limited there is, nevertheless, a vast field for it that has not yet been invaded.

A comparatively recent example of the practical application of electric resistance heating or welding is that of rivet heating, which from all indications will soon largely supplant the fire method of rivet heating. The rivet is heated by placing it between copper electrodes in the form Of blocks. A heavy current is then passed through the rivet, and in a few seconds the proper heat is attained. The riveting and handling of the rivets is otherwise the same as with the fire method of heating. Some of the advantages of the electric rivet heater are: Better control of heat resulting in fewer rivets burned; the rivet is more uniformly chances for elimination of smoke and dirt results in better efficiency of workmen; and last, but not least, the fire hazard is greatly minimized.

***Thermite Welding** . In the year 1894 it was found that the ignition of finely powdered aluminum, mixed with metallic oxides, produced an exceedingly high temperature because of the rapid oxidation of the aluminum. These facts were turned to practical account by Dr. H. Goldschmidt, who welded two iron bars by molten iron produced by the process to which the name of "thermite" is now commonly applied. This process has been wonderfully successful and has been extensively used especially for welding members of large cross-sections and for heated, thus reducing the*

ineffective riveting; the emergency repairs on certain classes of work. Thermite welding is sometimes.

Electric arc welding called a casting process, since it requires a mold around the parts to be joined.

Gas Welding . *The oxy-hydrogen blowpipe was first used about the year 1820 chiefly for producing limelight. It was also used in some important industrial applications, one of which was the fusion of platinum. In the latter part of the nineteenth century this process came into extensive use for lead burning, or welding. About the same time it was discovered that by using oxy-acetylene a much higher flame temperature could be secured, which together with improved regulation of heat control, led to the extremely rapid use and extension of the oxy-acetylene torch to the welding and cutting of iron and steel, and other metals to a lesser degree. While other gases have been used in place of acetylene, the oxy-acetylene flame is by far the most widely used.*

Today the oxy-acetylene welding and cutting process is used in practically all of the metal using industries.

Electric Arc Welding. *Electric arc welding is commercially the most recent and newest process of any form of welding.*

Benardos and Slavianoff are generally credited with the discovery of the possibilities of the carbon arc and metallic arc, respectively, for the welding of metals. The carbon arc process was the first one to be used for welding metals, and was first used, on a small scale,30 years ago. This form of arc welding is sometimes called the Benardos process. Not long after the carbon arc process was demonstrated by Benardos, Slavianoff demonstrated the possibilities of the metallic arc process, but it was not until comparatively recent years that either was used to any appreciable commercial extent.

After the first discovery of the more or less vague possibilities of electric arc welding the progress in the development of the art was extremely slow, due to the fact that it was only with great difficulty that the work of development could be carried on.

There were several reasons for the existence of such a condition, most important of which was the fact that the men who first conceived and worked to develop and improve the welding art were apparently versed only in one branch or phase of that science.

Understand the process of shielded Metal Arc Welding. An electric arc is formed at the tip of the welding rod when a current passes across an air gap and continues through the grounded metal which is being welded. Here are some of the terms and their descriptions used in this article:

Welding machine . *This is the term used to describe the machine which converts 120-240 volt AC electricity to welding voltage, typically 40-70 volts*

AC, but also a range of DC voltages. It generally consists of a large, heavy transformer, a voltage regulator circuit, an internal cooling fan, and an amperage range selector. The term welder applies to the person doing the welding. A welding machine requires a welder to operate it.

Leads, or Welding leads. These are the insulated copper conductors which carry the high amperage, low voltage electricity to the work piece that is being welded.

Rod holder , or stinger is the device on the end of the lead that holds the electrode, which the person welding uses to accomplish the welding task.

Ground and ground clamp . This is the lead that grounds, or completes the electrical circuit, and specifically, the clamp that is attached to the work to allow the electricity to pass through the metal being welded.

Amperage, or amps . This is an electrical term, used to describe the electrical current supplied to the electrode.

DC and reverse polarity . This is a different configuration used in welding with an arc/electrode system, which offers more versatility, especially in overhead welding applications and for use welding certain alloys that do not weld easily with AC voltages. The welding machine that produces this current has a rectifier circuit or has the current supplied by a generator, and is much more expensive than a typical AC welder.

Electrodes. There are many specialized welding electrodes, used for specific alloys and types of metals, such as cast or malleable iron, stainless or chromium steel, aluminum, and tempered or high carbon steels. A typical electrode consists of the wire rod in the center covered with a special coating (flux)which burns as the arc is maintained, consuming oxygen and producing carbon dioxide in the weld area to prevent the base metal from oxidizing or burning away in the arc flame during the welding process. Here are some common electrodes and their uses:

E6011 electrodes are a mild steel electrode with a cellulose fiber coating. The first two numbers in the electrode identification is the tensile strength, measured in pounds per square inch times 1,000. Here, the yield of the electrode would be 60,000 PSI.

E6010 electrodes are a reverse polarity electrode, commonly used for welding steam and water pipes, and are particularly useful for overhead welding, since the metal holds its position while in a liquid state, being drawn into the molten weld pool by the flow of the direct current from the electrode to the workpiece.

Other specific purpose E60XX electrodes are available, but since E6011s are considered a standard, general purpose rod, and E6010s are considered the standard for reverse polarity DC welding, they will not be covered in detail in this article.

E7018 electrodes are low hydro-gen-flux coated steel rods, with a high yield tensile strength of 70,000 PSI. These are often used in assembling structural steel used in the construction industry, and in other applications where a strong filler material and higher strength weld is required. Note that, although these rods provide greater strength, they are less forgiving in respect to achieving a clean, high-grade weld at incorrect amperages and with dirty (rusted, painted, or galvanized) steels. These electrodes are called low hydrogen due to every attempt to lower the hydrogen content. These electrodes must be stored in an oven with a temperature between 250°F and 300°F. This temperature is above the water boiling point of 212°F at sea level. This temperature keeps the moisture (dew)(H_2O) in the air from collecting in the flux.

Nickel, Cast-alloy, Ni-Rod electrodes . These are special rods made for welding cast, ductile, or malleable iron, and have more yield, to allow for the expansion and contraction of the iron material being welded.

Dissimilar metals rods . These rods are made from a special alloy and give better results when welding tempered, hardened or alloyed steels.

Aluminum rods. These are a more recent technology and allow arc welding aluminum with a conventional welder, rather than using a special gas-shielded wire feed welder like a MIG (metal, inert gas) or TIG (tungsten, inert gas) welding machine, often referred to as heliac welding, since helium was the gas used to shield the arc flame while welding. The official names created by the American Welding Society for this arc type welding are Shielded

Metal Arc Welding (stick), Gas Tungsten Arc Welding (tig)and Gas Metal Arc Welding (mig).

Electrode sizes. Electrodes come in a variety of sizes, measured by the diameter of the metal center of each rod. For mild steel rods, a diameter range of 1/16 inch to 3/8 inch is available, and the size used is determined by *the amperage of the welder, and the thickness of the material being welded. Each rod performs best at a given amperage range. Selecting the correct amperage range for a given size rod will depend on the base material and the desired penetration, so specific amperages will only be covered for the welding described further in this article.*

Safety equipment . *A critical part of welding safely is having, and knowing how to use, the correct safety equipment for the job. Here are some typical items that are required for welding safely.*

Welding shield (hood) . *This is the mask which is worn to protect the person welding from the bright flash of the arc, and from sparks being thrown during welding. Standard arc welding lenses are tinted very darkly, since exposure to the arc flash can cause flash burns to the retina of the eye. A level 10 darkness is the minimum for arc welding. Welding hoods with a flip up lens was once preferred, as the dark lens can be lifted up, and a separate clear glass lens will protect the welder from bits of slag while the weld is chipped. The newer self darkening welding shields are the most desirable welding shield now sold. These welding shield lens are very light colored for grinding and torch cutting. When an arc is struck the automatic self darkening lens will change to a preset #10 shade. Even newer on the market are the variable shade automatic self darkening lens.*

Welding gloves . *These are special, insulated leather gloves that reach about 6 inches above the wrists, and protect the hands and lower arms of the welder (the person welding). They also provide limited protection from accidental shock if the person welding comes into contact with the electrode accidentally.*

Welding leathers . *This is an apron like leather jacket that covers the shoulders and chest of the welder, used for overhead work where sparks might ignite the welder's clothing, or cause burns.*

Work boots . *The person welding should wear at least a 6 inch lace-up type boot to prevent sparks and hot slag from burning his feet. These boots should have insulating soles made from a material which does not melt or burn easily.*

Learn the steps for creating a successful weld. Welding is more than dragging a welding rod across a piece of steel and gluing it to another one. The process begins with properly fitting and securing the work pieces, or metal to be welded, together. For thicker pieces, you may want to grind a bevel so subsequent beads can be placed in the groove to fill it completely with a solid weld. Here are the basic steps for completing a simple weld.

Strike the arc . This is the process of creating an electric arc between the electrode and the workpiece. If the electrode simply allows the current to pass directly into the grounded work piece, there will not be enough heat produced to melt and fuse metal together.

Move the arc to create a bead. The bead is the metal from the melting electrode flowing together with molten metal from the base metal to fill the space between the pieces being joined by welding.

Shape the weld bead . This is done by weaving the arc back and forth across the weld path either in a zig zag or figure 8 motion so the metal spreads to the width that you want your finished weld bead to be.

Chip and brush the weld between passes . Each time you complete a pass, or trip from one end to the other of your weld, you need to remove the slag, or the melted electrode flux material, from the surface of the weld bead so only clean molten metal will be filling the weld on the subsequent pass.

Gather the tools and materials you will need to begin welding. This means the welding machine, electrodes, cables and clamps, and the metal to be welded.

Set up a safe work area, preferably with a table constructed of steel or other non-flammable material. For practice, a few pieces of mild steel, at least 3/16 inch thick will work.

Prepare the metal to be welded . If the metal consists of two pieces that are to be joined in the welding process, you may need to prep, or weld prep them, by grinding a beveled edge on the sides that are to be joined. This allows for sufficient penetration of the weld arc to melt both sides to a molten state so the filler metal bonds through the sectional thickness of the metal. At the least, you should remove any paint, grease, rust, or other contaminants so you are working with a clean pool of molten metal as you weld.

Attach clamps to hold your metal pieces together, if need. Locking type pliers, "C" clamps, a vice, or spring loader clamps will usually work. For special projects, you may find you will have to adapt different techniques to secure the work pieces until they are joined.

Attach the ground clamp to the larger piece of stock that is being welded. Make sure there is a clean location so that the electrical circuit can be completed with minimal resistance at the ground location. Again, rust or paint will interfere with the grounding of your work piece, making it difficult to create an arc when you begin welding.

Select the correct rod and amperage range for the work you are attempting. As an example, 1/4 inch plate steel can be welded effectively using an E6011, 1/8 inch electrode, at between 80-100 amps. Place the electrode in the electrode holder (henceforth referred to as the stinger) making sure the conductive material of the stinger clamp is on the clean metal at the end of the electrode.

Turn on your welding machine . You should hear a humming sound from the transformer. The sound of the cooling fan running may or may not be heard. Some welding machine fans only operate when the machine requires cooling. If you do not, you may need to check the circuit that is supplying your power, and the breakers in the panel box. Welding machines require a considerable amount of power to operate, often a special circuit rated at 60 amps or more at 240 volts.

Hold the stinger in your dominant hand by the insulated handle, with the rod in a position so that striking the tip of it against the plate you are welding will be as natural a movement as possible. Hold your welding shield up just high enough so you can see to move the electrode to within a few inches of the workpiece, ready to flip in down to protect your eyes. You may want to practice tapping the electrode against the weld metal to get the feel of it before turning the power on, but never strike an electric arc without protecting your eyes.

Select the point where you wish to begin your weld. Position the tip of the rod close to it, then drop the welding hood into place. You want to tap the tip of the electrode against the metal to complete the electrical circuit, then instantaneously pull it back a little bit, to create an electric arc between the electrode tip and the metal being welded. Another way to strike an arc is like striking a match. This arc gap, or airspace, creates a great deal of resistance in the electrical circuit, which is what produces the arc flame or plasma and heat needed to liquefy the electrode and the metal adjacent to the weld area.

Strike the electrode against the surface of the metal, pulling it back slightly when you see an electric arc occur. This takes a great deal of practice, since different electrode diameters and welding amperages require a different gap between the tip of the electrode and the work piece, but if you can hold the *gap steady, a continuous electric arc will occur from the electrode to the work piece. Typically, the arc gap should be no greater than the electrode diameter. Practice steadying the arc by holding the electrode about 1/8 to 3/16 of an inch from the work piece, then begin moving along the path you want to weld. As you move the electrode, the metal will be melting away, filling the pool of molten metal and building your weld.*

Practice traveling across the path of your weld with the electrode until you can keep a consistent arc, moving at a consistent speed, and in line with the path you want to weld. When you have mastered controlling the arc, you will begin to practice laying, or building up the weld bead. This is the deposit of metal that joins the two pieces that you are welding together. The technique you use for laying your bead will depend on the width of the gap (if there is one) you are filling, and the depth you want the weld bead to penetrate. The slower you move the electrode, the deeper the weld will go into the metal work pieces, and for making a wide path, the more you zig zag or weave the electrode's tip, the wider the bead you will lay up.

***Keep the arc established** as you move along the weld you are making. If the electrode grounds to the metal and becomes stuck, jerk the stinger to break the rod free either from the stinger clamp or the weld metal. If the arc is lost because you move the electrode too far from the metal's surface, stop the process and clean the slag from the spot you are welding so when you re-strike the arc to continue, there will be no slag in the weld area to contaminate the new weld you are beginning from the place the arc was lost or broken. Never lay a new bead over existing slag, as this material will melt in the arc plasma and bubble through the new layer of metal you are placing, resulting in a weak and dirty weld.*

***Practice moving the electrode in a sweeping motion to create a wider bead.** This will allow you to fill more of the weld in a single pass, leaving a cleaner and more sound weld. The electrode is moved in a sideways motion as it is drawn along the weld path, either in a zig-zag, curved, or figure eight motion.*

***Adjust your welder's output amperage to suit the material you are welding and the desired penetration of the arc.** If you find the finished weld bead is pitted, with deep cratering at the bead's edges, or the adjacent metal is simply melted or burned away, reduce the amperage incrementally until the condition is corrected. If, on the other hand, you have difficulty striking or* *maintaining an arc, you may need to increase the amperage.*

Clean your finished weld. *After you have finished welding, you may want to remove the slag and clean up your weld, either to allow paint to bond better, or simply for cosmetic reasons. Chip off the slag and wire brush the weld to remove any foreign material and remaining slag. If the surface needs to be flat to allow fitting the piece you have welded to another piece, use an angle grinder to remove the top, or high portion of the bead. A clean weld, particularly after grinding flat, is easier to examine to see if pitting, puddling, or other defects have occurred while welding.*

Paint your weld with a suitable rust-preventative primer to protect it from corrosion. *Freshly welded metal will corrode rather quickly if exposed to the elements, since the actual base metal is exposed directly to moisture.*

Before you get started welding - and this is very important - get comfortable, use two hands to hold the handle, and brace yourself in any and every possible way to make sure you are in a comfortable position. This is one of the biggest secrets journeyman welders use all of the time. I was originally thought to weld with one hand and all that means for you is less control. Control is the biggest factor when it comes to welding technique!

Electrode sticking

Getting the arc started is easier said than done. In the beginning, you will find that the rod sticks and the flux will likely chip off and ruin a small part of the rod. It's almost unavoidable and you have to take it like learning to ride a bike. Even the best welders have this happen now and then. To help with striking the arc, if you are wearing the proper welding gloves that are dry, you can lean the rod on the other hand almost like a pool stick and strike it. Once the arc starts you put that hand with the other one on the welding handle.

To fix any chipped flux, take a piece of scrap metal and strike an arc holding the rod about a ¼ of an inch away from the metal till it burns to a complete and undamaged part of the rod. This is the only time you will want a long arc like that. After you have a good rod again, I find removing the rod from the holder and scraping it on something to remove the used flux helps with restarting the arc later. It gives the metal inside the rod a good contact to strike an arc with much less effort.

There are a few ways to move the rod and see the puddle of metal in the crater. In stick welding, you normally drag the rod in MIG and in TIG you push the torch with only a few exceptions. If you are right handed, you will point the rod toward your left side and strike it like a match to the right or bounce it lightly off of the metal. It may take a few tries to get the arc started but once you do you want to keep the arc as short as possible while scraping the metal with the edge of the rod on your movements. If your arc is too far from the metal the shield of gas protecting the arc from oxygen will not be able to do so. This will result in a porous weld that looks like Swiss cheese on the inside. That means the weld is not very strong. The welding rod should stay ahead of the puddle to avoid trapping slag in the weld. Your eyes should also focus mostly on the size if the puddle behind the rod to determine the size of the weld.

With most welding rods you can use the same techniques in most positions, but there are certain patterns welders favor for good reason. The most common techniques for a E6011 are:

1) Whipping the rod, a moving it back and forth motion.
2) Circles to fuse the metal in a circular motion.
3) Weaving a side to side motion (for wider welds).

Then there are others, like myself, who have no real pattern. We only see the size of the puddle and adapt to what the fill needs of the joint are. When I was learning, circles worked for me, or half-circles. It all comes down to what you feel comfortable doing and gives you the best results! In the beginning, pick a pattern and stick to it until the weld is good. The rest will come in time!

Basic welding guidelines are that for 6010 and 6011 you whip the rod. For low hydrogen rods like a 7018 or 7024 you do circles or just hold a steady position and wait for the rod to fill the gap. With low hydrogen rods or any rods that have a lot of flux you want to avoid whipping because the flux that protects it also can cause problems by getting trapped in the weld. Technically this is called a slag inclusion. Don't forget that the weld should twice the width of the rod.

Welding technique for thinner metals should be whipped; it keeps the rod from burning a hole in the metal. For metals slightly thicker than the rod, do circles or whip depending how wide you want the weld. Finally, for thicker metals after you complete the first pass using one of the above techniques, you should start weaving using a side-to-side motion and concentrating on holding onto the sides for the most part. The best example I can give on weaving technique is to actually count out loud the number 1001 while holding on one side and then weaving over to the other side and then repeating. It produces a very consistent pattern which produces a solid weld with excellent penetration. Welding in different position changes what you do and how hot you set your machine. Flat or 1G /1F as it is called is the easiest. Horizontal or 2G /2F is a bit harder.

Safety first

Gas welding or cutting shall not be done in or near rooms or locations where flammable liquids or vapors, lint, dust or loose combustible stocks are located or arranged so that sparks or hot metal from the welding or cutting operations may cause ignition or explosion of such materials.

2. When such welding or cutting must be done above or within 20 feet of combustible construction or material, non-combustible shields shall be interposed to protect such materials and persons from sparks and hot metal or oxide.

3. One or more fire extinguishers of a suitable approved type shall be kept at the location where the welding or cutting is being done. All extinguishers must have current service tags, seal, etc.

4. When welding or cutting is done above or within 20 feet of combustible construction or material, a "fire watch" shall be on hand to make use of fire extinguishing equipment. 5. A fire watch shall be maintained in the area of welding or cutting operations for at least 1 hour after completion of such operations to detect and extinguish possible smoldering fires.

6. All compressed gas cylinders in service or in storage shall be adequately secured to prevent falling or being knocked over.

7. Oxygen and fuel gas cylinders and acetylene generators shall be placed far enough away from the welding position that they will not be unduly heated by radiation from heated materials, by sparks or slag, or by misdirection of the torch flame.

8. Cylinders, valves, regulators, hose and other apparatus and fittings containing or using oxygen shall be kept free from oil or grease. Oxygen cylinders, apparatus and fittings shall not be handled with oily hands, gloves or greasy materials.

9. Fuel gas shall never be used from cylinders through torches or other devices equipped with shut-off valves without reducing the pressure through a suitable regulator attached to the cylinder valve or manifold.

10. Oxygen shall never be used from a cylinder or cylinder manifold unless a pressureregulating device intended for use with oxygen, and so marked, is provided.

11. The user shall not transfer gases from one cylinder to another or mix gases in a cylinder.

12. Acetylene gas shall not be brought in contact with unalloyed copper except in a blowpipe

13. PREPARING THE CLEAN CONTAINER FOR WELDING OR CUTTING--INERT GAS TREATMENT

a. General. Inert gas may be used as a supplement to any of the cleaning methods and as an alternative to the water filling treatment. If sufficient inert gas is mixed with flammable gases and vapors, the mixture will come non-flammable. A continuous flow of steam may also be used. The steam will reduce the air concentration and make the air flammable gas mixture too lean to burn. Permissible inert gases include carbon dioxide and nitrogen.

b. Carbon Dioxide and Nitrogen.

(1) When carbon dioxide is used, a minimum concentration of 50 percent is required, except when the falmmable vapor is principally hydrogen, carbon monoxide, or acetylene. In these cases, a minimum concentration of 80 percent carbon dioxide is required. Carbon dioxide is heavier than air, and during welding or cutting operations will tend to remain in containers having a top opening.

(2) When nitrogen is used, the concentrations should be at least 10 percent greater than those specified for carbon dioxide.

(3) Do not use carbon monoxide.

c. Procedure. The procedure for inert gas, carbon dioxide, or nitrogen treatment is as follows:

(1) Close all openings in the container except the filling connection and vent. Use damp wood flour or similar material for sealing cracks or other damaged sections.

(2) Position the container so that the spot to be welded or cut is on top. Then fill it with as much water as possible.

(3) Calculate the volume of the space above the water level and add enough inert gas to meet the minimum concentration for nonflammability. This will usually require a greater volume of gas than the calculated minimum, since the inert gas may tend to flow out of the vent after displacing only part of the previously contained gases or vapors.

(4) Introduce the inert gas, carbon dioxide, or nitrogen from the cylinder through the container drain at about 5 psi (34.5 kPa). If the drain connection cannot be used, introduce the inert gas through the filling opening or vent. Extend the hose to the bottom of the container or to the water level so that the flammable gases are forced out of the container.

(5) If using solid carbon dioxide, crush and distribute it evenly over the greatest possible area to obtain a rapid formation of gas.

d. Precautions When Using Carbon Dioxide. Avoid bodily contact with solid carbon dioxide, which may produce "burns". Avoid breathing large amounts of carbon dioxide since it may act as a respiratory stimulant, and, in sufficient quantities, can act as an asphyxiant.

e. Inert Gas Concentration. Determine whether enough inert gas is present using a combustible gas indicator instrument. The inert gas concentration must be maintained during the entire welding or cutting operation. Take steps to maintain a high inert gas concentration during the entire welding or cutting operation by one of the following methods:
(1) If gas is supplied from cylinders, continue to pass the gas into the container.

(2) If carbon dioxide is used in solid form, add small amounts of crushed solid carbon dioxide at intervals to generate more carbon dioxide gas.

WARNING

Welding polyurethane foam-filled parts can produce toxic gases. Welding should not be attempted on parts filled with polyurethane foam. If repair by welding is necessary, the foam must be removed from the heat-affected area, including the residue, prior to welding.

a. General. Welding polyurethane foam filled parts is a hazardous procedure. The hazard to the worker is due to the toxic gases generated by the thermal breakdown of the polyurethane foam. The gases that evolve from the burning foam depend on the amount of oxygen available. Combustion products of polyurethane foam in a clean, hot fire with adequate oxygen available are carbon dioxide, water vapor, and varying amounts of nitrogen oxides, carbon monoxide, and traces of hydrogen cyanide. Thermal decomposition of polyurethanes associated with restricted amounts of oxygen as in the case of many welding operations results in different gases being produced. There are increased amounts of carbon monoxide, various aldehydes, isocyanates and cyanides, and small amounts of phosgene, all of which have varying degrees of toxicity.

b. Safety Precautions.

(1) It is strongly recommended that welding on polyurethane foam filled parts not be performed. If repair is necessary, the foam must be removed from the heat affected zone. In addition, all residue must be cleaned from the metal prior to welding.

(2) Several assemblies of the M113 and M113A1 family of vehicles should not be welded prior to removal of polyurethane foam and thorough cleaning.

Gas Shield Arc Welding Precautions

When any of the welding processes are used, the shielded from the air in order to obtain a high molten puddle of metal should be quality weld deposit. In shielded metal arc welding, shielding from the air is accomplished by gases produced by the disintegration of the coating in the arc. With gas shielded arc welding, shielding from the air is accomplished by surrounding the arc area with a localized gaseous atmosphere throughout the welding operation at the molten puddle area.

Gas shielded arc welding processes have certain dangers associated with them. These hazards, which are either peculiar to or increased by gas shielded arc welding, include arc gases, radiant energy, radioactivity from thoriated tungsten electrodes, and metal fumes.

Protective measures
a. Gases.

(1) Ozone. Ozone concentration increases with the type of electrodes used, amperage, extension of arc tine, and increased argon flow. If welding is carried out in confined spaces and poorly ventilated areas, the ozone concentration may increase to harmful levels. The exposure level to ozone is reduced through good welding practices and properly designed ventilation systems.

(2) Nitrogen Oxides. Natural ventilation may be sufficient to reduce the

hazard of exposure to nitrogen oxides during welding operations, provided. Nitrogen oxide concentrations will be very high when performing gas tungsten-arc cutting of stainless steel using a 90 percent nitrogen-10 percent argon mixture. Also, high concentrations have been found during experimental use of nitrogen as a shield gas. Good industrial hygiene practices dictate that mechanical ventilation.

(3) Carbon Dioxide and Carbon Monoxide. Carbon dioxide is disassociated by the heat of the arc to form carbon monoxide. The hazard from inhalation of these gases will be minimal if ventilation requirements.

WARNING

The vapors from some chlorinated solvents (e.g., carbon tetrachloride, trichloroethylene, and perchloroethylene) break down under the ultra-violet radiation of an electric arc and forma toxic gas. Avoid welding where such vapors are present. Furthermore, these solvents vaporize easily and prolonged inhalation of the vapor can be hazardous. These organic vapors should be removed from the work area before welding is begun. Ventilation shall be provided for control of fumes and vapors in the work area.

(4) Vapors of Chlorinated Solvents. Ultraviolet radiation from the welding or cutting arc can decompose the vapors of chlorinated hydrocarbons, such as perchloroethylene, carbon tetrachloride, and trichloroethylene, to form highly toxic substances. Eye, nose, and throat irritation can result when the welder is exposed to these substances. Sources of the vapors can be wiping rags, vapor de-greasers, or open containers of the solvent. Since this de-composition can occur even at a considerable distance from the arc, the source of the chlorinated solvents should be located so that no solvent vapor will reach the welding or cutting area.

b. Radiant Energy. Electric arcs, as well as gas flames, produce ultraviolet and infrared rays which have a harmful effect on the eyes and skin upon continued or repeated exposure. The usual effect of ultraviolet is to "sunburn" the surface of the eye, which is painful and disabling but generally temporary. Ultraviolet radiation may also produce the same effects on the skin as a severe sunburn. The production of ultraviolet radiation doubles when gas-shielded arc welding is performed. Infrared radiation has the effect of heating the tissue with which it comes in contact. Therefore, if the heat is not sufficient to cause an ordinary thermal burn, the exposure is minimal. Leather and WoOl clothing is preferable to cotton clothing during gas-shielded arc welding. Cotton clothing disintegrates in one day to two weeks, presumably because of the high ultraviolet radiation from arc welding and cutting.

c. Radioactivity from Thoriated Tungsten Electrodes. Gas tungsten-arc welding using these electrodes may be employed with no significant hazard to the welder or other room occupants. Generally, special ventilation or protective equipment other than that specified is not needed for protection from exposure hazards associated with welding with thoriated tungsten electrodes.

d. Metal Fumes. The physiological response from exposure to metal fumes varies depending upon the metal being welded. Ventilation and personal protective equipment requirements shall be employed to prevent hazardous exposure

Welding machines

a. When electric generators powered by internal combustion engines are used inside buildings or in confined areas, the engine exhaust must be conducted to the outside atmosphere.

b. Check the welding equipment to make sur the electrode connections and the insulation on holders and cables are in good condition. All checking should be done with the machine off or unplugged. All serious trouble should be investigated by a trained electrician.

c. Motor-generator welding machines feature complete separation of the primary power and the welding circuit since the generator is mechanically connected to the electric rotor. A rotor-generator type arc welding machine must have a power ground on the machine. Metal frames and cases of motor generators must be grounded since the high voltage from the main line does come into the case. Stray current may cause a severe shock to the operator if he should contact the machine and a good ground. d. In transformer and rectifier type welding machines, the metal frame and cases must be grounded to the earth. The work terminal of the welding machine should not be grounded to the earth.

e. Phases of a three-phase power line must be accurately identified when paralleling transformer welding machines to ensure that the machines are on the same phase and in phase with one another. To check, connect the work leads together and measure the voltage between the electrode holders of the two machines. This voltage should be practically zero. If it is double the normal open circuit voltage, it means that either the primary or secondary connections are reversed. If the voltage is approximately 1-1/2 times the normal open circuit voltage it means that the machines are connected to different phases of the three phase power line. Corrections must be made before welding begins.

f. When large weldments, like ships, buildings, or structural parts are involved, it is normal to have the work terminal of many welding machines connected to it. It is important that the machines be connected to the proper phase and have the same polarity. Check by measuring the voltage between the electrode holders of the different machines as mentioned above. The situation can also occur with respect to direct current power sources when they are connected to a common weldment. If one machine is connected for straight polarity and one for reverse polarity, the voltage between the electrode holders will be double the normal open circuit voltage. Precautions should be taken to see that all machines are of the same polarity when connected to a common weldment.

g. Do not operate the polarity switch while the machine is operating under welding current load. Consequent arcing at the switch will damage the contact surfaces and the flash may burn the person operating the switch.

h. Do not operate the rotary switch for current settings while the machine is operating under welding current load. Severe burning of the switch contact surfaces will result. Operate the rotary switch while the machine is idling. i. Disconnect the welding machines from the power supply when they are left unattended.

j. The welding electrode holders must be connected to machines with flexible cables for welding application. Use only insulated electrode holders and cables. There can be no splices in the electrode cable within 10 feet (3 meters) of the electrode holder. Splices, if used in work or electrode leads, must be insulated. Wear dry protective covering on hands and body.

k. Partially used electrodes should be removed from the holders when not in use. A place will be provided to hang up or lay down the holder where it will not come in contact with persons or conducting objects.

l. The work clamp must be securely attached to the work before the start of the welding operation.
m. Locate welding machines where they have adequate ventilation and

ventilation ports are not obstructed.

Electric circuits

a. A shock hazard is associated with all electrical equipment, including extension lights, electric hand tools, and all types of electrically powered machinery. Ordinary household voltage (115 V) is higher than the output voltage of a conventional arc welding machine.

b. Although the ac and dc open circuit voltages are low compared to voltages used for lighting circuits and motor driven shop tools, these voltages can cause severe shock, particularly in hot weather when the welder is sweating. Consequently, the precautions listed below should always be observed.
(1) Check the welding equipment to make certain that electrode connections and insulation on holders and cables are in good condition.

(2) Keep hands and body insulated from both the work and the metal electrode holder. Avoid standing on wet floors or coming in contact with grounded surfaces.

(3) Perform all welding operations within the rated capacity of the welding cables. Excessive heating will impair the insulation and damage the cable leads.

Fuel gas cylinders

a. Although the most familiar fuel gas used for cutting and welding is acetylene, propane, natural gas, and propylene are also used. Store these fuel gas cylinders in a specified, well-ventilated area or outdoors, and in a vertical condition.
b. Any cylinders must have their caps on, and cylinders, either filled or empty, should have the valve closed.

c. Care must be taken to protect the valve from damage or deterioration. The major hazard of compressed gas is the possibility of sudden release of the gas by removal or breaking off of the valve. Escaping gas which is under high pressure will cause the cylinder to act as a rocket, smashing into people and property. Escaping fuel gas can also be a fire or explosion hazard.

d. In a fire situation there are special precautions that should be taken for acetylene cylinders. All acetylene cylinders are equipped with one or more safety relief devices filled with a low melting point metal. This fusible metal melts at about the killing point of water (212°F or 100°C). If fire occurs on or near an acetylene cylinder the fuse plug will melt. The escaping acetylene

may be ignited and will burn with a roaring sound. Immediately evacuate all people from the area. It is difficult to put out such a fire. The best action is to put water on the cylinder to keep it cool and to keep all other acetylene cylinders in the area cool. Attempt to remove the burning cylinder from close proximity to other cylinders, from flammable or materials, or from combustible buildings. It is best to allow the gas to burn rather than to allow acetylene to escape, mix with air, and possibly explode.
acetylene hazardous

e. If the fire on a cylinder is a small flame around the hose connection, the valve stem, or the fuse plug, try to put it out as quickly as possible. A wet glove, wet heavy cloth, or mud slapped on the flame will frequently extinguish it. Thoroughly wetting the gloves and clothing will help protect the person approaching the cylinder. Avoid getting in line with the fuse plug which might melt at any time.

f. Oxygen cylinders should be stored separately from fuel gas cylinders and separately from combustible materials. Store cylinders in cool, well-ventilated areas. The temperature of the cylinder should never be allowed to exceed 130°F (54°C).

g. When cylinders are empty they should be marked empty and the valves must be closed to prohibit contamination from entering.

h. When the gas cylinders are in use a regulator is attached and the cylinder should be secured to prevent falling by means of chains or clamps.

i. Cylinders for portable apparatuses should be securely mounted in specially designed cylinder trucks.

j. Cylinders should be handled with respect. They should not be dropped or struck. They should never be used as rollers. Hammers or wrenches should not be used to open cylinder valves that are fitted with hand wheels. They should never be moved by electromagnetic cranes. They should never be in an electric circuit so that the welding current could pass through them. An arc strike on a cylinder will damage the cylinder causing possible fracture, requiring the cylinder to be condemned and discarded from service.

Hoses

a. Do not allow hoses to come in contact with oil or grease. These will penetrate and deteriorate the rubber and constitute a hazard with oxygen.

b. Always protect hoses from being walked on or run over. Avoid kinks and tangles. Do not leave hoses where anyone can trip over them. This could result in personal injury, damaged connections, or cylinders being knocked over. Do not work with hoses over the shoulder, around the legs, or tied to the waist.

c. Protect hoses from hot slag, flying sparks, and open flames.

d. Never force hose connections that do not fit. Do not use white lead, oil, grease, or other pipe fitting compounds for connections on hose, torch, or other equipment. Never crimp hose to shut off gases.
e. Examine all hoses periodically for leaks by immersing them in water while under pressure. Do not use matches to check for leaks in acetylene hose. Repair leaks by cutting hose and inserting a brass splice. Do not use tape for mending. Replace hoses if necessary.

f. Make sure that hoses are securely attached to torches and regulators before using.

g. Do not use new or stored hose lengths without first blowing them out with compressed air to eliminate talc or accumulated foreign matter which might otherwise enter and clog the torch parts.

h. Only approved gas hoses for flame cutting or welding should be used with oxygen fuel gas equipment. Single lines, double vulcanized, or double multiple stranded lines are available.

i. The size of hose should be matched to the connectors, regulators, and torches.
j. In the United States, the color green is used for oxygen, red for acetylene or fuel gas, and black for inert gas or compressed air. The international standard calls for blue for oxygen and orange for fuel gas.

k. Connections on hoses are right-handed for inert gases and oxygen, and left-handed for fuel gases.

l. The nuts on fuel gas hoses are identified by a groove machined in the center of the nuts.

MAPP gas cylinders
a. MAPP gas is a mixture of stabilized methylacetylene and propadiene.
b. Store liquid MAPP gas around 70°F (21°C) and under 94 psig pressure.

c. Repair any leaks immediately. MAPP gas vaporizes when the valve is opened and is difficult to detect visually. However, MAPP gas has an obnoxious odor detectable at 100 parts per million, a concentration 1/340th *of its lower explosive limit in air. If repaired when detected, leaks pose little or no danger. However, if leaks are ignored, at very high concentrations (5000 parts per million and above) MAPP gas has an anesthetic effect.*

d. Proper clothing must be worn to prevent injury to personnel. Once released into the open air, liquid MAPP gas boils at -36 to -4°F (-54 to -20°C). This causes frost-like burns when the gas contacts the skin.

e. MAPP gas toxicity is rated very slight, but high concentrations (5000 part per million) may have an anesthetic affect.

f. MAPP gas has some advantages in safety which should be considered when choosing a process fuel gas, including the following:
(1) MAPP gas cylinders will not detonate when dented, dropped, or incinerated.

(2) MAPP gas can be used safely at the full cylinder pressure of 94 psig.
(3) Liquified fuel is insensitive to shock.
(4) Explosive limits of MAPP gas are low compared to acetylene.
(5) Leaks can be detected easily by the stron smell of MAPP gas.
(6) MAPP cylinders are easy to handle due to their light weight.
OXYGEN CYLINDERS

a. Always refer to oxygen by its full name and not by the word "air" alone.
b. Oxygen should never be used for "air" in any way.

WARNING

Oil or grease in the presence of oxygen will ignite violently, especially in an enclosed pressurized area.

c. Oxygen cylinders shall not be stored near highly combustible material, especially oil and grease; near reserve stocks of carbide and acetylene or other fuel gas cylinders, or any other substance likely to cause or accelerate fire; or in an acetylene generator compartment.

d. Oxygen cylinders stored in outside generator houses shall be separated from the generator or carbide storage rooms by a noncombustible partition having a fire resistance rating of at least 1 hour. The partition shall be without openings and shall be gas tight.

e. Oxygen cylinders in storage shall be separated from fuel gas cylinders or combustible materials (especially oil or grease) by a minimum distance of 20.0 ft (6.1 m) or by a noncombustible barrier at least 5.0 ft (1.5 m) high and having a fire-resistance rating of at least one-half hour.

f. Where a liquid oxygen system is to be used to supply gaseous oxygen for welding or cutting and a bulk storage system is used, it shall comply with the provisions of the Standard for Bulk Oxygen Systems at Consumer Sites, NFPA No. 566-1965, National Fire Protection Association.

g. When oxygen cylinders are in use or being roved, care must be taken to avoid dropping, knocking over, or striking the cylinders with heavy objects. Do not handle oxygen cylinders roughly.

h. All oxygen cylinders with leaky valves or safety fuse plugs and discs should be set aside and marked for the attention of the supplier. Do not tamper with or attempt to repair oxygen cylinder valves. Do not use a hammer or wrench to open the valves.

i. Before attaching the pressure regulators, open each oxygen cylinder valve for an instant to blow out dirt and foreign matter from the nozzle. Wipe off the connection seat with a clean cloth. Do not stand in front of the valve when opening it.

WARNING

Do not substitute oxygen for compressd air in pneumatic tools. Do not use oxygen to blow out pipe lines, test radiators, purge tanks or containers, or to "dust" clothing or work.

j. Open the oxygen cylinder valve slowly to prevent damage to regulator high pressure gage mechanism. Be sure that the regulator tension screw is released the before opening the valve. When not in use, the cylinder valve should be closed and the protecting caps screwed on to prevent damage to the valve.

k. When the oxygen cylinder is in use, open the valve to the full limit to prevent leakage around the valve stem.

l. Always use regulators on oxygen cylinders to reduce the cylinder pressure to a low working pressure. High cylinder pressure will burst the hose.

m. Never interchange oxygen regulators, hoses, or other apparatus with similar equipment intended for other gases.

Acetylene cylinders

If acetylene cylinders have been stored or transported horizontally (on their sides), stand cylinders vertically (upright) for 45 minutes prior to (before) use.

a. Always refer to acetylene by its full name and not by the word "gas" alone. Acetylene is very different from city or furnace gas. Acetylene is a compound of carbon and hydrogen, produced by the reaction of water and calcium carbide.

b. Acetylene cylinders must be handled with care to avoid damage to the valves or the safety fuse plug. The cylinders must be stored upright in a well ventilated, well protected, dry location at least 20 ft from highly combustible materials such as oil, paint, or excelsior. Valve protection caps must always be in place, hand-tight, except when cylinders are in use. Do not store the cylinders near radiators, furnaces, or in any are with above normal temperatures. In tropical climates, care must be taken not to store acetylene in areas where the temperature is in excess of 137°F (58°C). Heat will increase the pressure, which may cause the safety fuse plug in the cylinder to blow out. Storage areas should be located away from elevators, gangways, or other places where there is danger of cylinders being knocked over or damaged by falling objects.

c. A suitable truck, chain, or strap must be used to prevent cylinders from falling or being knocked over while in use. Cylinders should be kept at a safe distance from the welding operation so there will be little possibility of sparks, hot slag, or flames reaching them. They should be kept away from radiators, piping systems, layout tables, etc., which may be used for grounding electrical circuits. Nonsparking tools should be used when changing fittings on cylinders of flammable gases.

d. Never use acetylene without reducing the pressure with a suitable pressure reducing regulator. Never use acetylene at pressures in excess of 15 psi.

e. Before attaching the pressure regulators, open each acetylene cylinder valve for an instant to blow dirt out of the nozzles. Wipe off the connection seat with a clean cloth. Do not stand in front of valves when opening them.

f. Outlet valves which have become clogged with ice should be thawed with warm water. Do not use scalding water or an open flame.

g. Be sure the regulator tension screw is released before opening the cylinder valve. Always open the valve slowly to avoid strain on the regulator gage which records the cylinder pressure. Do not open the valve more than one and one-half turns. Usually, one-half turn is sufficient. Always use the special T-wrench provided for the acetylene cylinder valve. Leave this wrench on the stem of the valve tile the cylinder is in use so the acetylene can be quickly *turned off in an emergency.*

h. Acetylene is a highly combustible fuel gas and great care should be taken to keep sparks, flames, and heat away from the cylinders. Never open an acetylene cylinder valve near other welding or cutting work.

i. Never test for an acetylene leak with an open flame. Test all joints with soapy water. Should a leak occur around the valve stem of the cylinder, close the valve and tighten the packing nut. Cylinders leaking around the safety fuse plug should be taken outdoors, away from all fires and sparks, and the valve opened slightly to permit the contents to escape.

j. If an acetylene cylinder should catch fire, it can usually be extinguished with a wet blanket. A burlap bag wet with calcium chloride solution is effective for such an emergency. If these fail, spray a stream of water on the cylinder to keep it cool.

k. Never interchange acetylene regulators, hose, or other apparatus with similar equipment intended for oxygen.
l. Always turn the acetylene cylinder so the valve outlet will point away from the oxygen cylinder.

m. When returning empty cylinders, see that the valves are closed to prevent escape of residual acetylene or acetone solvent. Screw on protecting caps.

n. Make sure that all gas apparatus shows UL or FM approval, is installed properly, and is in good working condition.

o. Handle all compressed gas with extreme care. Keep cylinder caps on when not in use.

p. Make sure that all compressed gas cylinders are secured to the wall or other structural supports. Keep acetylene cylinders in the vertical condition.

q. Store compressed gas cylinders in a safe place with good ventilation. Acetylene cylinders and oxygen cylinders should be kept apart.

r. Never use acetylene at a pressure in excess of 15 psi (103.4 kPa). Higher pressure can cause an explosion.

s. Acetylene is nontoxic; however, it is an anesthetic and if present in great enough concentrations, is an asphyxiant and can produce suffocation.

Welding in confined spaces

a. A confined space is intended to mean a relatively small or restricted space such as a tank, boiler, pressure vessel, or small compartment of a ship or tank.

b. When welding or cutting is being performed in any confined space, the gas cylinders and welding machines shall be left on the outside. Before operations are started, heavy portable equipment mounted on wheels shall be securely blocked to prevent accidental movement.

c. Where a welder must enter a confined space through a manhole or other all opening, means will be provided for quickly removing him in case of emergency. When safety belts and life lines are used for this purpose, they will be attached to the welder's body so that he cannot be jammed in a small exit opening. An attendant with a preplanned rescue procedure will be stationed outside to observe the welder at all times and be capable of putting rescue operations into effect.

d. When arc welding is suspended for any substantial period of time, such as during lunch or overnight, all electrodes will be removed from the holders with the holders carefully located so that accidental contact cannot occur. The welding machines will be disconnected from the power source.

e. In order to eliminate the possibility of gas escaping through leaks or improperly closed valves when gas welding or cutting, the gas and oxygen supply valves will be closed, the regulators released, the gas and oxygen lines bled, and the valves on the torch shut off when the equipment will not be used for a substantial period of time. Where practical, the torch and hose will also be removed from the confined space.

f. After welding operations are completed, the welder will mark the hot metal or provide some other means of warning other workers

Ventilation for General Welding and Cutting.

(1) General. Mechanical ventilation shall be provided when welding or cutting is done on metals not covered in subparagraphs i through p of this section, and under the following conditions:

(a) In a space of less than 10,000 cu ft (284 cu m) per welder.
(b) In a roan having a ceiling height of less than 16 ft (5 m).

(c) In confined spaces or where the welding space contains partitions, balconies, or other structural barriers to the extent that they significantly obstruct cross ventilation.

(2) Minimum rate. Ventilation shall be at the minimum rate of 200 cu ft per

minute (57 cu m) per welder, except where local exhaust heeds, as in paragraph 2-4 g below, or airline respirators approved by the US Bureau of Mines, National Institute of Occupational Safety and Health, or other government-approved testing agency, are used. When welding with rods larger than 3/16 in. (0.48 cm) in diameter, the ventilation shall be higher.

Personal protective equipment

a. General. The electric arc is a very powerful source of light, including visible, ultraviolet, and infrared. Protective clothing and equipment must be worn during all welding operations. During all oxyacetylene welding and cutting processes, operators must use safety goggles to protect the eyes from heat, glare, and flying fragments of hot metals. During all electric welding processes, operators must use safety goggles and a hand shield or helmet equipped with a suitable filter glass to protect against the intense ultraviolet and infrared rays. When others are in the vicinity of the electric welding processes, the area must be screened so the arc cannot be seen either directly or by reflection from glass or metal.

b. Helmets and Shields.
(1) Welding arcs are lights. They contain intensely brilliant

a proportion of ultraviolet light which may cause eye damage. For this reason, the arc should never be viewed with the naked eye within a distance of 50.0 ft (15.2 m). The brilliance and exact spectrum, and therefore the danger of the light, depends on the welding process, the metals in the arc, the arc atmosphere, the length of the arc, and the welding current. Operators, fitters, and those working nearby need protection against arc radiation. The intensity of the light from the arc increases with increasing current and arc voltage. Arc radiation, like all light radiation, decreases with the square of the distance. Those processes that produce smoke surrounding the arc have a less bright arc since the smoke acts as a filter. The spectrum of the welding arc is similar to that of the sun. Exposure of the skin and eyes to the arc is the same as exposure to the sun.

(2) Being closest, the welder needs a helmet to protect his eyes and face from harmful light and particles of hot metal. The welding helmet is generally constructed of a pressed fiber insulating material. It has an adjustable headband that makes it usable by persons with different head sizes. To minimize reflection and glare produced by the intense light, the helmet is dull black in color. It fits over the head and can be swung upward when not

welding. The chief advantage of the helmet is that it leaves both hands free, making it possible to hold the work and weld at the same time.

(3) The hand-held shield provides the same protection as the helmet, but is held in position by the handle. This type of shield is frequently used by an observer or a person who welds for a short period of time.

(4) The protective welding helmet has lens holders used to insert the cover glass and the filter glass or plate. Standard size for the filter plate is 2 x 4-1/4 in. (50 x 108 mm). In some helmets lens holders open or flip upwards. Lenses are designed to prevent flash burns and eye damage by absorption of the infrared and ultraviolet rays produced by the arc. The filter glasses or plates come in various optical densities to filter out various light intensities, depending on the welding process, type of base metal, and the welding current. The color of the lens, usually green, blue, or brown, is an added protection against the intensity of white light or glare. Colored lenses make it possible to clearly see the metal and weld. Table 2-1 lists the proper filter shades to be used. A magnifier lens placed behind the filter glass is sometimes used to provide clear vision.

To prevent injury to personnel, extreme caution should be exercised when using any types of welding equipment. Injury can result from fire, explosions, electric shock, or harmful agents. Both the general and specific safety precautions listed below must be strictly observed by workers who weld or cut metals.

b. Do not permit unauthorized persons to use welding or cutting equipment.

c. Do not weld in a building with wooden floors, unless the floors are protected from hot metal by means of fire resistant fabric, sand, or other fireproof material. Be sure that hot sparks or hot metal will not fall on the operator or on any welding equipment components.
d. Remove all flammable material, such as cotton, oil, gasoline, etc., from the vicinity of welding.

e. Before welding or cutting, warm those in close proximity who are not protected to wear proper clothing or goggles.

f. Remove any assembled parts from the component being welded that may become warped or otherwise damaged by the welding process.

g. Do not leave hot rejected electrode stubs, steel scrap, or tools on the floor or around the welding equipment. Accidents and/or fires may occur.

h. Keep a suitable fire extinguisher nearby at all times. Ensure the fire extinguisher is in operable condition.

i. Mark all hot metal after welding operations are completed. Soapstone is commonly used for this purpose.

Fire hazards

a. Fire prevention and protection is the responsibility of welders, cutters, and supervisors. Approximately six percent of the fires in industrial plants are caused by cutting and welding which has been done primarily with portable equipment or in areas not specifically designated for such work. The elaboration of basic precautions to be taken for fire prevention during welding or cutting is found in the Standard for Fire Prevention in Use of Cutting and Welding Processes, National Fire Protection Association Standard 51B, 1962. Some of the basic precautions for fire prevention in welding or cutting work are given below.

b. During the welding and cutting operations, sparks and molten spatter are formal which sometimes fly considerable distances. Sparks have also fallen through cracks, pipe holes, or other small openings in floors and partitions, starting fires in other areas which temporarily may go unnoticed. For these reasons, welding or cutting should not be done near flammable materials unless every precaution is taken to prevent ignition.

c. Hot pieces of base metal may come in contact with combustible materials and start fires. Fires and explosions have also been caused when heat is transmitted through walls of containers to flammable atmospheres or to combustibles within containers. Anything that is combustible or flammable is susceptible to ignition by cutting and welding.

d. When welding or cutting parts of vehicles, the oil pan, gasoline tank, and other parts of the vehicle are considered fire hazards and must be removed or effectively shielded from sparks, slag, and molten metal.

e. Whenever possible, flammable attached to or near equipment materials requiring

welding, brazing, or cutting will be removed. If removal is not practical, a suitable shield of heat resistant material should be used to protect the flammable material. Fire extinguishing equipment, for any type of fire that may be encountered, must be present.

Vertical up rod angle

Vertical welding means you can be welding from bottom to top or top to bottom. There is vertical down which is pretty easy but is not very strong. Vertical down works great on thinner metals. Then there us vertical up. The direction of a vertical weld changes the amount of penetration and overall strength. Think of a water hose aimed at a mound of dirt. If you spray from the top down you only succeed in carving into the mound a little. But if you spray from the bottom up you gouge a deeper grove. Vertical up is a very strong weld but much harder to do.

Most welding shops require a vertical up or 3G welding certification or better.

Finally there is overhead or 4G / 4F welding. Literally meaning welding over your head. It's just like welding flat or horizontal, but with way more sparks hitting you!

Typically, setting your machine for flat will allow you to weld flat, horizontal, vertical down, and overhead depending on the rod type. Vertical up usually required less amperage. What happens with vertical up is that when you weld, the rods arc gouges a crater into the metal and then fills the joint with filler metal. That is why the weld is so strong even though the setting are lower.

Welding position basics rod angles are as follows and are just a guideline. These are not set in stone!

Flat or 1G you drag the rod between 10 to 30 degrees in the direction of your movement. Use either a whip or circular motion.

Horizontal or 2G you point the rod upward at 45 degrees and drag it toward your direction of travel with a side tilt of 10 to 30 degrees. Also use a whip or circular motion.

Vertical up or 3G point the rod up at 45 degrees and use a tight side to side motion or a weave depending on the width of the weld you need.

Over head or 4G is the same as flat or 1G except the rod is pointed up.

Apart from the use of rivets, screws, etc., metal is commonly joined by soldering, brazing, or welding, three groups of processes that have one thing in common the use of heat to fuse either the metals themselves or an alloy which is interposed to consolidate the joint. The word " solder " is derived through the French from a Latin word meaning "solid."

Soldering may be "soft " or "hard." Soft-soldering uses lead-tin alloys which are easily melted in a bunseti as flame or with a hot iron or bit; while hard-soldering employs a silver-copper alloy, to melt which a mouth blowpipe at least is necessary. Brazing is hard-soldering with spelter (brass), and a forge or a heavy blowlamp or a powerful blowpipe must be employed to provide the heat.

Welding is a fusion process which in the past was almost entirely confined to wrought-iron and steel, these metals possessing the property of being able to "fuse together" to an extent unknown in the case of any other metals.

The old time blacksmith's process of welding is to heat the iron or steel until the surface of the metal becomes pasty, and then to bring the two pieces into intimate contact by hammering on the anvil. Of late years the welding of iron, steel, copper and some other metals has been rendered possible by the use of certain electrical and chemical methods and most important of all by the use of the oxy-acetylene, blow-pipe, the process being known as "fusion welding" or "autogenous soldering," the word autogenous implying that the process is complete in itself and independent of the use of any extraneous substance such as solder. The thermite process, of which so much has been heard, and which is brief and later, is the fusion welding of iron and steel by means of the intense heat produced

by the combustion of a special chemical compound.(Dura-aluminum 'Al' + Iron oxide 'FeO2'). Perhaps the oldest of the autogenous soldering processes is "lead-burning,"in which the flame of an airo-hydrogen blowpipe is brought to bear upon the lead, the joint being fed with a strip of the same metal.

Soft-soldering is an operation that the beginner will not find nearly so difficult as hard-soldering or brazing, and although the strength of joints made by it is not nearly equal to that produced by the methods named, it fills a useful place within its scope. It is purely a surface union " that is, the solder adheres to the faces in contact in much the same manner as an adhesive sticks to metal ; but with the assistance of fluxes, the contact is made so intimate that some force is necessary to break the joint. Soft-soldering also of use where would simply the ruin or destroy" the metals, as in the cases of lead, poor-quality brass, Pewter, tin, zinc, and in tin plate and galvanized iron.

In silver-soldering and brazing, the silver or spelter that fuses to form the joint alloys itself so intimately with the copper or brass that it actually becomes part of the piece itself, and for all practical purposes cannot be distinguished from it. But soft-soldering is not always inferior to hard-soldering. Indeed,

the surface nature of the soldering often constitutes the strongest joints of all are produced by fusion.

A solder should melt at a slightly lower temperature than the metals which it unites, and should possess the quality of alloying with the two surfaces, thus effecting a sound and true metallic joint. Ordinary soft solders are lead-tin alloys, and the larger the proportion of lead the commoner is the solder said to be. At an extreme is plumber's solder, consisting of 2 parts of lead to 1 part of tin, and, at the other, the best blowpipe soft solder, which contains 2 parts of tin to only 1 part of lead.

In the ordinary way\a "coarse " or "common "

solder is 2 parts of lead to 1 part of tin ; a " fine " or " medium " solder, 1 part of lead to 1 part of tin ; and a "very fine " or "best " solder, 1 part of lead to 2 parts of tin.

Eutectio Alloys . Liead - tin solders are eutectic alloys" that is, they are examples of the phenomenon of a combination of two metals melting at a temperature lower than one of them would if melted separately.

Thus, lead melts at about 328" C., and tin at about 232" C., yet will show that the "commonest "solder mentioned fuses at 303" C., and the "best " at 175" C.

Melting
10 90 303"
20 80 278"
80 70 255"
40 60 230"
60 50 205"
('.(" 40 187"
63 37 175"
70 30 185"
80 20 198"
90 10 215"

Hardness of solders. According to the before-mentioned authority, the hardness of various lead-tin alloys by Brineli's method, by which a steel cone *is forced into the metal. The results he obtained are as follow :*

Lead

...

100 90 80 70 CO 50 40 34 S3 32 30 20 10 0 Tin

...

0 10 20 30 40 50 60 66 67 08 70 80 90 100

*Hardness 3'9 lO'l 12-16 14-5 15'8 IS'O 14'6 107 16'4 14'6 16'8 15*2 18'3 4'1*

These results, show that the hardest alloy is the one with 66" (about 2 parts)of tin and 34" (about 1 part)of lead, which also is the one having the lowest melting point of all the lead-tin alloys. The results also show that tin is slightly harder than lead.

Compositions of soft solders *." As already shown, solders vary in the ability to fuse, according to their composition and the choice should be determined by the nature of the work and the properties of the metal to be soldered.*

Should a solder be used of too high a melting-point, the metal will itself be fused before the solder begins to flow.

A point to be particularly observed is that the introduction of a foreign substance into the solder for example, the addition of a little zinc to a pot of "very fine" solder will utterly spoil it and render it un-workable.

To remove zinc from solder, melt the solder in a pot, take it off the fire and stir in powdered sulfur or brimstone until the whole is of the consistency of wet sand. Replace the pot on the fire and melt, but do not stir the contents. The sulfur and zinc will rise to the surface and form into a cake.

Now take the pot off the fire and carefully remove the cake without breaking by employing two pieces of hoop iron with bent ends.

It is false economy to use a rough solder for fine work on the score of cheapness, since more solder is required for a given job on account of the rough particles of solder clinging to the work; moreover, the rough appearance of the soldering may completely spoil the job.

Making solder strips, wire, tears, etc.

Only clean, pure tin and pure lead should be employed. The lead is first melted and then the tin added. When all is melted, place a piece of resin on the molten metal to act as a flux, and after well stirring, the solder is made into strips by pouring from a ladle. Solder should not be poured into sand. It may be poured into strips on an oiled sheet of black iron, preferably corrugated to accommodate the strips.

In the absence of a corrugated iron sheet, some workers use a ladle resembling a large spoon with a hole about 1/4 inch in diameter near the end. To form the strips, get a ladle full of solder, place it on a flat iron sheet; then, tilting the ladle to allow the solder to flow over the hoe, quickly draw the ladle across the sheet.

Strip of solder should thus be formed, and the thickness of the strip may be varied by increasing or decreasing the diameter of the hole in the ladle. A button of solder usually forms at one or both ends of the strip, and this excess should be melted off the strips by just dipping the ends into the molten solder

in the pot. Solder wire is very handy for small work, and can be made in the following way: Koll a sheet of stiff writing or drawing paper into a conical form, rather broad in comparison with its length; make a ring of stiff wire to hold it in, attaching a suitable handle to the ring. The point of the cone should first of all be cut off to leave an orifice of the size required.

It should then be filled with molten solder, and held above a pail of cold water, and the stream of solder flowing from the cone will solidify as it runs and form the wire.

If held a little higher, so that the stream of solder breaks into drops before striking the water, it will form handy elongated "tears" of metal; when it is held still higher, each drop forms a thin concave cup or shell, and each of these forms will be found to have its own unique traits in blowpipe work.

The method adopted for granulating tin man's

solder, which is very rarely called tor, is as follows:
Place a piece of wood, well greased, over a tub containing water, and by gently pouring the molten alloy from a distance in a small stream on to the greased board, the metal is broken up into a large number of very fine shots, which run off the board into the water and gently dried. Making Solder from Pewter. " This alloy is composed of variable proportions of tin and lead, the average composition being about 4 parts of lead to 1 part of tin. If old pewter is to be utilized for making solder, tin will have to be added to the molten pewter. Thus, to convert 5 Ib. of average pewter.

Making Coarse Solder from Composition Piping. Good composition piping is made of nearly all tin, or an alloy of tin and lead, in which the former metal is in excess, and formerly was much used by plumbers in the making of coarse solder, as the material consisted of odd pieces of small value. As, however, a great deal of composition tubing is made out of old metals of which lead, tin, antimony, arsenic, and zinc form the alloy, it is not advisable to introduce it into solder.

*Should it be done, the melting point of the solder would be raised, and in applying it to the lead to be joined together, would probably partly melt it. Neither do the metals named alloy in a thorough manner, but partake more of the nature of a mixture in which the soldering, brazing and welding constituents partly separate when making the joints, and some, especially zinc, slow as small bright lumps on the surface. Joints wiped with what is usually called "poisoned metal" are difficult to make, almost invariably leak when on water service pipes, and are dirty grey, instead of **bright and clean.** **The zinc could be removed from the mixture by the method already given.***

Combined solder and flux.

This consisted of equal parts of lead and tin made into fine tubing and afterwards filled with flux having resin as a base.

" Tin-ol" is a paste made of finely powdered solder and a special flux, and there is also "Tin-ol wire " having a core of flux.

A "magic" solder, sold in most hardware stores, consists of the above tubular flux-filled solder of such low melting point that it can be fused in the flame of a lighted match.

Soft solders that melt in boiling water.

The following soft solders melt at a temperature lower than that of boiling water: 1 part tin, 1 part lead, and 2 parts bismuth, melting point about 200" F. ; 8 parts lead, 4 parts tin, 15 parts bismuth, and 3 parts cadmium, melting point 140" to 150" F. ; 6 parts lead, 7 parts bismuth, and 1 part cadmium, melting point about 180" F. To ensure the alloys melting at the temperatures stated, the metals of which they are formed should be free from impurities, and care should be taken to prevent oxidation while making the alloys.

When uniting the metals, that having the highest melting point should be melted first, with a layer of resin over it, the other metals being added in the order of their melting points. The alloy should then be well stirred with a wooden stick, and poured quickly into molds.

Re-melting and overheating solder.

After solder has been re-melted a number of times or has been overheated, its content of tin will be reduced, and the solder will become poorer and coarser. The tin melts earlier than the lead and, being the lighter of the two, floats over it, and is thus fully exposed to the air, the oxidizing effect of which on heated, molten metal is extremely active. The oxidized tin forms a dross, from which most of the tin may, however, be recovered by melting it with powdered charcoal, which combines with the oxygen and frees the tin. The addition of a little fresh tin is desirable. This will make the solder re-usable or clean everything and start from scratch with fresh solder.

Often times, this is the better of the two solutions when soldering critical components or very dedicate work.

Soldering aluminum

It is well known to those accustomed to the art of soldering that there is no solder which operates with aluminum in the same way that ordinary solders operate with tin-plate, copper, brass, etc. Aluminum soldering presents so many difficulties that it has been thought desirable to devote a separate chapter to the

subject.

There is more than one reason for the difficulty encountered. Aluminum does not alloy readily with solders at temperatures as low as other metals require; and, secondly, aluminum alloys with lead solders only with great difficulty, and with but a small proportion of lead at that. Consequently, lead solders are not suitable for aluminum. Another and even more serious reason is in respect to the refractory oxide which forms at soldering temperatures, and which is undoubtedly responsible for most of the trouble.

The soldering of aluminum is one of the most debated subjects in metal working. Almost as soon as aluminum was prepared on a large scale, it was discovered that the ordinary solders and fluxes did not answer with it. Either pure tin or pure zinc will wet aluminum, and can, therefore, be used as solder for it; experience shows that the tin soon falls apart, while zinc by itself is brittle and discolors badly. The failure of tin is due to the fact that it forms with the aluminum an alloy that is decomposed by the action of the oxygen present in the air.

Although aluminum is popularly supposed to be non-oxidizable, really the surface is covered with a very thin film of oxide, which prevents solder from alloying with the metal. Aluminum when heated rapidly oxidizes. It is customary to scrape the metal before and during the soldering; and although some workers say that it is useless to scrape before soldering because oxidation immediately starts again, it is obvious that a thin film is more easily penetrated than a thick one. Often it answers to scrape with the copper bit during the soldering, previously rubbing off the oxide with emery cloth. The work should, if possible, be backed with asbestos, to keep up heat in the metal. To discover whether the surface is thoroughly tinned, wipe off lightly, and the un-tinned parts will then soon become apparent. If the oxide is not scraped off beforehand, it will probably mix with the solder and form a scum, which will make a neat flow difficult. Scum should be lightly removed with an old knife blade. It is essential to "tin " every part to be joined, as the solder will not take on any spot that has not been rubbed in some way, unless previously coated.

Solders for Aluminum.

Hundreds of aluminum solders have been invented, naturally all claimed to be strong and durable, the alloys containing various metals, such as aluminum, antimony, bismuth, cadmium, chromium, copper, lead, manganese, silver, tin, and zinc.

Many of the best solders for aluminum contain a small proportion of phosphor tin. A molten alloy containing phosphorus placed on aluminum tends to absorb oxygen from the impure film as well as the surrounding air.

Compositions o aluminum solders

To avoid loss of the more easily volatile of the metals by adopting the following precautions: The aluminum is melted first, the zinc is added in small pieces, then tin in small pieces, and lastly the phosphor tin.

Soldering, brazing and welding

Inasmuch as zinc alloys with aluminum more readily than does any of the common metals, solders that will readily tin, and aluminum generally contain zinc in varying proportions. The solders found most satisfactory contain zinc, tin, aluminum, and a very small proportion of phosphor tin; but they do not run very freely or fuse so readily as the ordinary tin and lead solders, and it is necessary to use a higher temperature, so high, in fact, that difficulty is found in using these solders with a soldering bit, and it is generally necessary to use a blowlamp.

While there is no solder that allows aluminum to be soldered with the facility and success experienced with other metals, is extensively used, and seems to have given as good results as any.

It consists of the following ingredients: Tin 29 parts zinc 11 parts, aluminum 1 part, and 5 per cent, phosphor tin 1 part practically the same as that given in the last line of the table. This solder has with- stood the test of time better than many of the patented solders, and can be used in jointing aluminum to aluminum to copper or brass, and without the use of a flux. In making the solder it is advisable to avoid loss of the more easily volatile of the metals. The aluminum should be melted first, then the zinc, tin, and phosphor tin in the order named.

When using phosphorus instead of phosphor tin in the making of aluminum solder, it will first be necessary to incorporate it with the tin, for which purpose take a length of 1-in. gas barrel, attach a screwed in the manipulation of the aluminum soldering bit owing to its lower melting temperature than the copper and nickel bits.

***The process of soldering aluminum** . The soldering of aluminum must be performed quickly to be satisfactory, as the metal, if not coated at the first attempt, may be injuriously affected.*

"Tinning" the parts required to be soldered first is another important factor ; also the distance of the overlap of the joints should not exceed more than J in.,so as to allow the solder to flow thoroughly through; it does not flow so readily as when soldering other metals.

In soldering large pieces, where the ordinary over- lap is not allowable, and where a butt joint would be weak, fit the pieces together as at a Solder always flows towards the hottest point.

This tendency enables one to direct its course under the blowpipe or blowlamp flame. A large flame should only be employed in "heating " up the part to be soldered on large and heavy work. With a small pointed flame directly on the solder and the parts on which it rests, the solder will flow quickly, and leave a smooth, even surface at completion.

Some aluminum solders now on the market are so hard that it is necessary to heat them and the work to redness before they melt. Sheet aluminum is easily warped by heat, and also contracts badly. If the solder is too high in melting point, the metal must also be brought to that point to cause proper union.

If a hole is being filled in, the body of the metal on heat. In cooling, the hole enlarges, tho, patch cm i tracts, and the solder also contracts; nicks result. The body of the work, if not exactly evenly made, will warp, which is fatal to engine ,and similar work. By a low-heat solder (melting point, about 700" F.)

Soldering aluminum to copper or brass *. Aluminum can be readily soldered to copper or brass with fine solder (2 parts of tin and 1 part of lead) : tin*

Aluminum Fitted Together for Soldering the metals, using stearin as flux previous to making. It is essential that both the tinning and soldering should be thoroughly done. Do not expect the solder to pull the joint together, but make sure that the joint is kept under slight pressure until the solder is hard, otherwise the joint will not be perfect.

Many workmen refuse to place any reliance in such joints.

Finally, it seems very likely that, at any rate as regards factory work, the use of solder on aluminum objects will be wholly discarded in the future in favor of fusion welding or autogenous soldering, in which process no alloy is interposed between the surfaces to be joined.

Wiping joints on lead pipes

Plumbers make joints in lead pipes with soft solder which, by means of cloths, they "wipe " to the shape. The difference between a copper-bit or blowpipe joint and a wiped joint.

Plumbers' Solder. As already stated, coarse, or plumbers' wiping solder, is made in the proportion of 2 of lead to 1 of block tin. Care must be taken that the lead is quite pure and free from any other metal, as zinc-adulterated solder will be difficult to use, and joints made with it on service pipes will not be sound.

In melting up scrap lead for making solder, only sheet lead should be used, as the lead used in the manufacture of sheets is much

purer and contains a greater proportion of pure pig lead. The scraps must be quite dry ; a damp piece dropped into a pot of molten metal

may cause a serious accident as the contents of the pot may be blown out.

To test the quality of solder when made, heat it as for wiping a joint; the correct temperature is deter- mined by dropping a small piece of newspaper into the pot, and if it quickly burns and catches alight the solder is right for

using. Next pour a small quantity on to a cold but dry stone or cement floor.
This, should have a few spots on the sin-funtiny-piece, and on the. under
should be bright nearly all over. Solder of this quality would, if properly
used, withstand any pressure without sweating. If the solder on the stone or
cement floor looks white on both sides, or has a few small bright spots on the
under side only, it is too and requires more tin. On no account should the
solder be heated to redness, as the tin rises to the top and quickly turns to
dross. If this should happen to solder that is being used for service pipes, it
should be rectified by adding more tin.

To purify a pot of "poisoned " solder (solder that contains zinc), melt, stir in
a handful of common sulphur or powdered brimstone until the mass is of the
constituency of wet sand, heat to the ordinary working temperature, and
carefully remove the crust that forms on the top, and the solder will then be
fit for use, except that a little tin must be added to it. The presence of zinc in
solder can be detected by the difficulty of forming joints, the metal falling
apart and working very lumpy, and the joints when finished having a dirty
grey appearance.

When plumber's solder is bought ready for use from the manufacturers, it is
usually in the form of casts of eight bars, weighing about 56 Ib. to the cast.
The best only should be used, as cheap solder is frequently the cause of much
trouble if used on high-pressure work, and joints made with it are never of
good appearance. To test"manufacturers' solder, wipe a joint with it, and if it
is of good quality it will work easily at a good heat, and when cleaned off
with tallow and a clean rag it should be well covered with bright spots.

Brass fittings should not be tinned by dipping into the solder pot, as brass
being an alloy of zinc and copper, the zinc may be melted into the pot with
disastrous results.

The flux used is tallow, no other flux answering the purpose so well, although
mutton fat has been used as a substitute. Plumbers often call tallow "touch,"
and they frequently use it in the form of tallow candles, the cotton wicks
coming in handy for packing spindles of taps and slides of gas pendants.

An excellent plumber's black, soil or smudge, can be bought in packets, and
requires only to be mixed with water before using. Ordinary black consists of
lamp- black, glue, and water. The black should be first mixed with water.

How to figure out any shop problem

When you are working in the shop or on any project. Even if you are following exact specifications and plans. A problem or two can arise and this is where you must be able to analyze and look at the problem from a different perspective.

This is where experience will come in handy. First literally take a break and have a cup of coffee, eat something, take some deep breaths. This will help your mind and then get up and move around or look at the problem from the complete opposite direction (literally) and slow down for a moment and a solution should be present if you apply some real critical thinking and know that any problem does have a solution.

This may take some time, trial and error, etc. But you will find a workable solution to your problem. I learned this simply but effective technique from my Grandfather. Who would have a smoke break and cup of coffee. Then go over in his "mind's eye".

Turning over in his mind the object for example and welded joint on an hinge plate.

Then see if it works actually and then apply it. It may take some extra effort to resolve a problem especially if it has not been done before, or you have not had experience with it.

But once you have accomplished figuring out a problem and fixing it correctly that will be an amazing reward in itself.

Another good technique is to get some paper and a pencil or pen and doodle, re-design, and make a solution for your shop problem. If you are working on a special project this may take some time and you could end up bringing it home and still working on a solution.

Do not be afraid to try new concepts or apply new materials to your work. In the next chapter I will go over what is given in 'tolerances' for certain materials, metals, etc.

Scientists and engineers, routinely use tables for materials that have tolerances as a safety switch to safeguard themselves when designing new structures, and equipment. However, you do not have to adhere to these same 'limits' as theses scientists and engineers.

This will give you an advantage when creating small art forms, or other structures, that no scientist or engineer would put into plans. Examples of these are; upside down pyramid structures-balanced on the apex point.

Welded to a metal base. An inverted pyramid is visually striking and would not be designed 'normally' without some type of supporting structures. However, for artistic purposes that have no danger of failing due to loads imposed on them. This can be permitted and then you can laugh as the same scientists and engineers figure out the load and stress of the materials used after you display it.

Another popular type of art exhibits are the hanging (usually large) metal art display's by very thin steel cable(s). These should be more or less tested before display to the public-if the artwork is to be anywhere above the heads of the on looking public.

Welding glossary

ACETONE: *A flammable, volatile liquid used in acetylene cylinders to dissolve and stabilize acetylene under high pressure.*

ACETYLENE: *A highly combustible gas composed of carbon and hydrogen. Used as a fuel gas in the oxyacetylene welding process.*

ACTUAL THROAT: *See THROAT OF FILLET WELD.*

AIR-ACETYLENE : *A low temperature flare produced by burning acetylene with air instead of oxygen.*

AIR-ARC CUTTING: *An arc cutting process in which metals to be cut are melted by the heat of the carbon arc. ALLOY: A mixture with metallic properties composed of two or more elements, of which at least one is a metal.*

ALTERNATING CURRENT: *An electric current that reverses its direction at regularly recurring intervals.*

AMMETER: *An instrument for measuring electrical current in amperes by an indicator activated by the movement of a coil in a magnetic field or by the longitudinal expansion of a wire carrying the current.*

ANNEALING: *A comprehensive term used to describe the heating and cooling cycle of steel in the solid state. The term annealing usually implies relatively slow cooling. In annealing, the temperature of the operation, the rate of heating and cooling, and the time the metal is held at heat depend upon the composition, shape, and size of the steel product being treated, and the purpose of the treatment. The more important purposes for which steel is* annealed are as follows: to remove stresses; to induce softness; to alter ductility, toughness, electric, magnetic, or other physical and mechanical properties; to change the crystalline structure; to remove gases; and to produce a definite micro-structure.

ARC BLOW: The deflection of an electric arc from its normal path because of magnetic forces.

ARC BRAZING: A brazing process wherein the heat is obtained from an electric arc formed between the base metal and an electrode, or between two electrodes.

ARC CUTTING: A group of cutting processes in which the cutting of metals is accomplished by melting with the heat of an arc between the electrode and the base metal. See CARBON-ARC CUTTING, METAL-ARC CUTTING, ARC-OXYGEN CUTTING, AND AIR-ARC CUTTING.

ARC LENGTH: The distance between the tip of the electrode and the weld puddle.

ARC-OXYGEN CUTTING: An oxygen-cutting process used to sever metals by a chemical reaction of oxygen with a base metal at elevated temperatures.

ARC VOLTAGE: The voltage across the welding arc.

ARC WELDING: A group of welding processes in which fusion is obtained by heating with an electric arc or arcs, with or without the use of filler metal.

AS WELDED : The condition of weld metal, welded joints, and weldments after welding and prior to any subsequent thermal, mechanical, or chemical treatments.

ATOMIC HYDROGEN WELDING: An arc welding process in which fusion is obtained by heating with an arc maintained between two metal electrodes in an atmosphere of hydrogen. Pressure and/or filler metal may or may not be used.

AUSTENITE: The non-magnetic form of iron characterized by a face-centered cubic lattice crystal structure. It is produced by heating steel above the upper critical temperature and has a high solid solubility for carbon and alloying elements.

AXIS OF A WELD: A line through the length of a weld, perpendicular to a cross section at its center of gravity.

BACK FIRE: The momentary burning back of a flame into the tip, followed *by a snap or pop, then immediate reappearance or burning out of the flame.*

BACK PASS: *A pass made to deposit a back weld.*

BACK UP *: In flash and upset welding, a locator used to transmit all or a portion of the upsetting force to the work pieces.*

BACK WELD: *A weld deposited at the back of a single groove weld.*

BACKHAND WELDING: *A welding technique in which the flame is directed towards the completed weld.*

BACKING STRIP: *A piece of material used to retain molten metal at the root of the weld and/or increase the thermal capacity of the joint so as to prevent excessive warping of the base metal.*

BACKING WELD: *A weld bead applied to the root of a single groove joint to assure complete root penetration.*

BACKSTEP: *A sequence in which weld bead increments are deposited in a direction opposite to the direction of progress.*

BARE ELECTRODE: *An arc welding electrode that has no coating other than that incidental to the drawing of the wire.*

BARE METAL-ARC WELDING: *An arc welding process in which fusion is obtained by heating with an unshielded arc between a bare or lightly coated electrode and the work. Pressure is not used and filler metal is obtained from the electrode.*

BASE METAL: *The metal to be welded or cut. In alloys, it is the metal present in the largest proportion.*

BEAD WELD: *A type of weld composed of one or more string or weave beads deposited on an unbroken surface.*

BEADING: *See STRING BEAD WELDING and WEAVE BEAD.*

BEVEL ANGLE: *The angle formed between the prepared edge of a member and a plane perpendicular to the surface of the member. BLACKSMITH WELDING: See FORGE WELDING.*

BLOCK BRAZING *: A brazing process in which bonding is produced by the heat obtained from heated blocks applied to the parts to be joined and by a nonferrous filler metal having a melting point above 800 °F (427 °C), but below that of the base metal. The filler metal is distributed in the joint by capillary attraction.*

BLOCK SEQUENCE: *A building up sequence of continuous multi-pass* *welds in which separated lengths of the weld are completely or partially built up before intervening lengths are deposited. See BUILDUP SEQUENCE.*

BLOW HOLE: *see GAS POCKET. BOND: The junction of the welding metal and the base metal.* **BOXING:** *The operation of continuing a fillet weld around a corner of a member as an extension of the principal weld.*

BRAZING: *A group of welding processes in which a groove, fillet, lap, or flange joint is bonded by using a nonferrous filler metal having a melting point above 800 °F (427 °C), but below that of the base metals. Filler metal is distributed in the joint by capillary attraction.*

BRAZE WELDING: *A method of welding by using a filler metal that liquefies above 450 °C (842 °F) and below the solid state of the base metals. Unlike brazing, in braze welding, the filler metal is not distributed in the joint by capillary action.*

BRIDGING: *A welding defect caused by poor penetration. A void at the root of the weld is spanned by weld metal.*

BUCKLING: *Distortion caused by the heat of a welding process.*

BUILDUP SEQUENCE *: The order in which the weld beads of a multi-pass weld are deposited with respect to the cross section of a joint. See BLOCK SEQUENCE.*

BUTT JOINT: *A joint between two work pieces in such a manner that the weld joining the parts is between the surface planes of both of the pieces joined.*

BUTT WELD: *A weld in a butt joint.*

BUTTER WELD: *A weld caused of one or more string or weave beads laid down on an unbroken surface to obtain desired properties or dimensions.*

CAPILLARY ATTRACTION: *The phenomenon by which adhesion between the molten filler metal and the base metals, together with surface tension of the molten filler metal, causes distribution of the filler metal between the properly fitted surfaces of the joint to be brazed.*

CARBIDE PRECIPITATION: *A condition occurring in austenitic stainless steel which contains carbon in a supersaturated solid solution. This condition is unstable. Agitation of the steel during welding causes the excess carbon in solution to precipitate. This effect is also called weld decay.*

CARBON-ARC CUTTING: *A process of cutting metals with the heat of an* *arc between a carbon electrode and the work.*

CARBON-ARC WELDING: *A welding process in which fusion is produced by an arc between a carbon electrode and the work. Pressure and/or filler metal and/or shielding may or may not be used.*

CARBURIZING FLAME: *An oxyacetylene flame in which there is an excess of acetylene. Also called excess acetylene or reducing flame.*

CASCADE SEQUENCE: *Subsequent beads are stopped short of a previous bead, giving a cascade effect.*

CASE HARDENING: *A process of surface hardening involving a change in the composition of the outer layer of an iron base alloy by inward diffusion from a gas or liquid, followed by appropriate thermal treatment. Typical hardening processes are carburizing, cyaniding, carbonitriding, and nitriding. CHAIN*

INTERMITTENT FILLET WELDS: *Two lines of intermittent fillet welds in a 'T' or lap joint in which the welds in one line are approximately opposite those in the other line.*

CHAMFERING: *The preparation of a welding contour, other than for a square groove weld, on the edge of a joint member.*

COALESCENCE: *The uniting or fusing of metals upon heating.*

COATED ELECTRODE: *An electrode having a flux applied externally by dipping, spraying, painting, or
burning, the
other similar methods. Upon coat produces a gas which
envelopes the arc.*

COMMUTATOR CONTROLLED WELDING: *The making of a number of spot or projection welds in which several electrodes, in simultaneous contact with the work, progressively function under the control of an electrical commutating device.*

COMPOSITE ELECTRODE: *A filler metal electrode used in arc welding, consisting of more than one metal component combined mechanically. It may or may not include materials that improve the properties of the weld, or stabilize the arc.*

COMPOSITE JOINT: *A joint in which both a thermal and mechanical process are used to unite the base metal parts.*

CONCAVITY: *The maximum perpendicular distance from the face of a concave weld to a line joining the toes.*

CONCURRENT HEATING: *Supplemental heat applied to a structure during the course of welding.* **CONE:** *The conical part of a gas flame next to the orifice of the tip.*

CONSUMABLE INSERT: *Pre-placed filler metal which is completely fused into the root of the joint and becomes part of the weld.*

CONVEXITY: *The maximum perpendicular distance from the face of a convex fillet weld to a line joining the toes.*

CORNER JOINT: *A joint between two members located approximately at right angles to each other in the form of an L.*

COVER GLASS: *A clear glass used in goggles, hand shields, and helmets to protect the filter glass from spattering material.*

COVERED ELECTRODE: *A metal electrode with a covering material which stabilizes the arc and improves the properties of the welding metal. The material may be an external wrapping of paper, asbestos, and other materials or a flux covering.*

CRACK: *A fracture type discontinuity characterized by a sharp tip and high ratio of length and width to opening displacement.* **CRATER:** *A depression at the termination of an arc weld.*

CRITICAL TEMPERATURE: *The transition temperature of a substance from one crystalline form to another.*

CURRENT DENSITY: *Amperes per square inch of the electrode cross sectional area.*

CUTTING TIP: *A gas torch tip especially adapted for cutting.*

CUTTING TORCH: *A device used in gas cutting for controlling the gases used for preheating and the oxygen used for cutting the metal*

CYLINDER: *A portable cylindrical container used for the storage of a compressed gas.*

DEFECT: *A discontinuity or discontinuities which, by nature or accumulated effect (for example, total crack length), render a part or product unable to meet the minimum applicable acceptance standards or specifications. This term designates reject ability.*

DEPOSITED METAL: *Filler metal that has been added during a welding* *operation.*

DEPOSITION EFFICIENCY: *The ratio of the weight of deposited metal to the net weight of electrodes consumed, exclusive of stubs.*

DEPTH OF FUSION: *The distance from the original surface of the base metal to that point at which fusion ceases in a welding operation.*

DIE: *a. Resistance Welding. A member, usually shaped to the work contour, used to clamp the parts being welded and conduct the welding current. b. Forge Welding. A device used in forge welding primarily to form the work while hot and apply the necessary pressure.*

DIE WELDING: *A forge welding process in which fusion is produced by heating in a furnace and by applying pressure by means of dies.*

DIP BRAZING: *A brazing process in which bonding is produced by heating in a molten chemical or metal bath and by using a nonferrous filler metal having a melting point above 800 °F (427 °C), but below that of the base metals. The filler metal is distributed in the joint by capillary attraction. When a metal bath is used, the bath provides the filler metal.*

DIRECT CURRENT ELECTRODE NEGATIVE (DCEN): *The arrangement of direct current arc welding leads in which the work is the positive pole and the electrode is the negative pole of the welding arc.*

DIRECT CURRENT ELECTRODE POSITIVE (DCEP): *The arrangement of direct current arc welding leads in which the work is the negative pole and the electrode is the positive pole of the welding arc.*

DISCONTINUITY: *An interruption of the typical structure of a weldment, such as lack of homogeneity in the mechanical, metallurgical, or physical characteristics of the material or weldment. A discontinuity is not necessarily a defect.*

DRAG: *The horizontal distance between the point of entrance and the point of exit of a cutting oxygen stream.*

DUCTILITY: *The property of a metal which allows it to be permanently deformed, in tension, before final rupture. Ductility is commonly evaluated by tensile testing in which the amunt of elongation and the reduction of area of the broken specimen, as compared to the original test specimen, are measured and calculated.* **DUTY CYCLE:** *The percentage of time during an arbitrary test period, usually 10 minutes, during which a power supply can be operated at its rated output without overloading.*

EDGE JOINT: *A joint between the edges of two or more parallel or nearly parallel members.* **EDGE PREPARATION:** *The contour prepared on the edge of a joint member for welding*

EFFECTIVE LENGTH OF WELD: *The length of weld throughout which the correctly proportioned cross section exits.*

ELECTRODE: *a. Metal-Arc. Filler metal in the form of a wire or rod, whether bare or covered, through which current is conducted between the electrode holder and the arc. b. Carbon-Arc. A carbon or graphite rod through which current is conducted between the electrode holder and the arc. c. Atomic Hydrogen. One of the two tungsten rods between the points of which the arc is maintained. d. Electrolytic Oxygen-Hydrogen Generation. The conductors by which current enters and leaves the water, which is decomposed by the passage of the current. e. Resistance Welding. The part or parts of a resistance welding machine through which the welding current and the pressure are applied directly to the work.*

ELECTRODE FORCE: *a. Dynamic. In spot, seam, and projection welding, the force (pounds) between the electrodes during the actual welding cycle. b. Theoretical. In spot, seam, and projection welding, the force, neglecting friction and inertia, available at the electrodes of a resistance welding machine by virtue of the initial force application and the theoretical mechanical advantage of the system. c. Static. In spot, seam, and projection welding, the force between the electrodes under welding conditions, but with no current flowing and no movement in the welding machine.*

ELECTRODE HOLDER: *A device used for mechanically holding the electrode and conducting current to it.*

ELECTRODE SKID: *The sliding of an electrode along the surface of the work during spot, seam, or projection welding.*

EMBOSSMENT: *A rise or protrusion from the surface of a metal.*
ETCHING: *A process of preparing metallic specimens and welds for macrographic or micrographic examination.*

FACE REINFORCEMENT: *Reinforcement of weld at the side of the joint from which welding was done.* FACE OF WELD: *The exposed surface of a weld, made by an arc or gas welding process, on the side from which welding was done.*

FAYING SURFACE: *That surface of a member that is in contact with* *another member to which it is joined.*

FERRITE: *The virtually pure form of iron existing below the lower critical temperature and characterized by a body-centered cubic lattice crystal structure. It is magnetic and has very slight solid solubility for carbon.*

FILLER METAL: *Metal to be added in making a weld.*

FILLET WELD: *A weld of approximately triangular cross section, as used in a lap joint, joining two surfaces at approximately right angles to each other.*

FILTER GLASS: *A colored glass used in goggles, helmets, and shields to exclude harmful light rays.*

FLAME CUTTING: *see OXYGEN CUTTING.*

FLAME GOUGING: *See OXYGEN GOUGING. FLAME HARDENING: A method for hardening a steel surface by heating with a gas flame followed by a rapid quench.*

FLAME SOFTENING: *A method for softening steel by heating with a gas flame followed by slow cooling.*

FLASH: *Metal and oxide expelled from a joint made by a resistance welding process.*

FLASH WELDING: *A resistance welding process in which fusion is produced, simultaneously over the entire area of abutting surfaces, by the heat obtained from resistance to the flow of current between application of substantially accompanied by expulsion of metal from the joint.*

FLASHBACK: *The burning of gases within the torch or beyond the torch in the hose, usually with a shrill, hissing sound.*

FLAT POSITION: *The position in which welding is performed from the upper side of the joint and two surfaces and by the*

pressure after heating is completed. Flashing is the face of the weld is approximately horizontal.

FILM BRAZING: *A process in which bonding is produced by heating with a molten nonferrous filler metal poured over the joint until the brazing temperature is attained. The filler metal is distributed in the joint by capillary attraction. See BRAZING.*

FLOW WELDING: *A process in which fusion is produced by heating with molten filler metal poured over the surfaces to be welded until the welding* *temperature is attained and the required filler metal has been added. The filler metal is not distributed in the joint by capillary attraction.*

FLUX: *A cleaning agent used to dissolve oxides, release trapped gases and slag, and to cleanse metals for welding, soldering, and brazing.*

FOREHAND WELDING: *A gas welding technique in which the flare is directed against the base metal ahead of the completed weld.*

FORGE WELDING: *A group of welding processes in which fusion is produced by heating in a forge or furnace and applying pressure or blows.*

FREE BEND TEST: *A method of testing weld specimens without the use of a guide.*

FULL FILLET WELD: *A fillet weld whose size is equal to the thickness of the thinner member joined.*

FURNACE BRAZING: *A process in which bonding is produced by the furnace heat and a nonferrous filler metal having a melting point above 800 °F (427 °C), but below that of the base metals. The filler metal is distributed in the joint by capillary attraction.*

FUSION: *A thorough and complete mixing between the two edges of the base metal to be joined or between the base metal and the filler metal added during welding.*

FUSION ZONE (FILLER PENETRATION): *The area of base metal melted as determined on the cross section of a weld.*

GAS CARBON-ARC WELDING: *An arc welding process in which fusion is produced by heating with an electric arc between a carbon electrode and the work. Shielding is obtained fran an inert gas such as helium or argon. Pressure and/or filler metal may or may not be used.*

GAS METAL-ARC (MIG) WELDING (GMAW): *An arc welding process in which fusion is produced by heating with an electric arc between a metal electrode and the work. Shielding is obtained from an inert gas such as helium or argon. Pressure and/or filler metal may or my not be used.*

GAS POCKET : *A weld cavity caused by the trapping of gases releasd by the metal when cooling.*

GAS TUNGSTEN-ARC (TIG) WELDING (GTAW) : *An arc welding process in which fusion is produced by heating with an electric arc between a tungsten electrode and the work while an inert gas forms around the weld* *area to prevent oxidation. No flux is used.*

GAS WELDING: *A process in which the welding heat is obtained from a gas flame.*

GLOBULAR TRANSFER (ARC WELDING) : *A type of metal transfer in which molten filler metal is transferred across the arc in large droplets.*

GOGGLES: *A device with colored lenses which protect the eyes from harmful radiation during welding and cutting operations.*

GROOVE: *The opening provided between two members to be joined by a groove weld.*

GROOVE ANGLE: *The total included angle of the groove between parts to be joined by a groove weld.*

GROOVE FACE: *That surface of a member included in the groove.*

GROOVE RADIUS: *The radius of a J or U groove.*

GROOVE WELD: *A weld made by depositing filler metal in a groove between two members to be joined.*

GROUND CONNECTION: *The connection of the work lead to the work.*

GROUND LEAD: *See WORK LEAD.*

GUIDED BEND TEST: *A bending test in which the test specimen is bent to a definite shape by means of a jig.*

HAMMER WELDING: *A forge welding process.*

HAND SHIELD: *A device used in arc welding to protect the face and neck. It is equipped with a filter glass lens and is designed to be held by hand.*

HARD FACING: *A particular form of surfacing in which a coating or cladding is applied to a surface for the main purpose of reducing wear or loss of material by abrasion, impact, erosion, galling, and cavitation.*

HARD SURFACING: *The application of a hard, wear-resistant alloy to the surface of a softer metal.*

HARDENING: *a. The heating and quenching of certain iron-base alloys from a temperature above the critical temperature range for the purpose of producing a hardness superior to that obtained when the alloy is not quenched. This term is usually restricted to the formation of martensite. b. Any process of increasing the hardness of metal by suitable treatment, usually involving heating and cooling.*

HEAT AFFECTED ZONE: *That portion of the base metal whose structure* or properties have been changed by the heat of welding or cutting.

HEAT TIME: *The duration of each current impulse in pulse welding.*

HEAT TREATMENT: *An operation or combination of operations involving the heating and cooling of a metal or an alloy in the solid state for the purpose of obtaining certain desirable conditions or properties. Heating and cooling for the sole purpose of mechanical working are excluded from the meaning of the definition.*

HEATING GATE: *The opening in a thermite mold through which the parts to be welded are preheated.*

HELMET: *A device used in arc welding to protect the face and neck. It is equipped with a filter glass and is designed to be worn on the head.*

HOLD TIME: *The time that pressure is maintained at the electrodes after the welding current has stopped.*

HORIZONTAL WELD: *A bead or butt welding process with its linear direction horizontal or inclined at an angle less than 45 degrees to the horizontal, and the parts welded being vertically or approximately vertically disposed.*

HORN: *The electrode holding arm of a resistance spot welding machine.*

HORN SPACING: *In a resistance welding machine, the unobstructed work clearance between horns or platens at right angles to the throat depth. This distance is measured with the horns parallel and horizontal at the end of the downstroke.*

HOT SHORT: *A condition which occurs when a metal is heated to that point, prior to melting, where all strength is lost but the shape is still maintained.*

HYDROGEN BRAZING: *A method of furnace brazing in a hydrogen atmosphere.*

HYDROMATIC WELDING: *See PRESSURE CONTROLLED WELDING.*
HYGROSCOPIC: *Readily absorbing and retaining moisture.*

IMPACT TEST: *A test in which one or more blows are suddenly applied to a specimen. The results are usually expressed in terms of energy absorbed or number of blows of a given intensity required to break the specimen.*

IMPREGNATED-TAPE METAL-ARC WELDING: *An arc welding process in which fusion is produced by heating with an electric arc between a* *metal electrode and the work. Shielding is obtained from decomposition of impregnated tape wrapped around the electrode as it is fed to the arc. Pressure is not used, and filler metal is obtained from the electrode.*

INDUCTION BRAZING: *A process in which bonding is produced by the heat obtained from the resistance of the work to the flow of induced electric current and by using a nonferrous filler metal having a melting point above 800 °F (427 °C), but below that of the base metals. The filler metal is distributed in the joint by capillary attraction.*

INDUCTION WELDING: *A process in which fusion is produced by heat obtained from resistance of the work to the flow of induced electric current, with or without the application of pressure.*

INERT GAS: *A gas which does not normally combine chemically with the base metal or filler metal.*

INTER-PASS TEMPERATURE: *In a multi-pass weld, the lowest temperature of the deposited weld meal before the next pass is started.*

JOINT: *The portion of a structure in which separate base metal parts are joined.*

JOINT PENETRATION: *The maximum depth a groove weld extends from its face into a joint, exclusive of reinforcement.*

KERF: *The space from which metal has been removed by a cutting process.*

LAP JOINT: *A joint between two overlapping members.*

LAYER: *A stratum of weld metal, consisting of one or more weld beads.*

LEG OF A FILLET WELD: *The distance from the root of the joint to the toe of the fillet weld.*

LIQUIDUS: *The lowest temperature at which a metal or an alloy is completely liquid.* **LOCAL PREHEATNG:** *Preheating a specific portion of a structure.*

LOCAL STRESS RELIEVING: *Stress relieving heat treatment of a specific portion of a structure.*

MANIFOLD: *A multiple header for connecting several cylinders to one or more torch supply lines.*

MARTENSITE: *Martensite is a microconstituent or structure in quenched steel characterized by an acicular or needle-like pattern on the surface of polish. It has the maximum hardness of any of the structures resulting from* *the decomposition products of austenite.*

MASH SEAM WELDING: *A seam weld made in a lap joint in which the thickness at the lap is reduced to approximately the thickness of one of the lapped joints by applying pressure while the metal is in a plastic state.*

MELTING POINT: *The temperature at which a metal begins to liquefy.*

MELTING RANGE: *The temperature range between solidus and liquidus.*

MELTING RATE: *The weight or length of electrode melted in a unit of time.*

METAL-ARC CUTTING: *The process of cutting metals by melting with the heat of the metal arc.*

METAL-ARC WELDING: *An arc welding process in which a metal electrode is held so that the heat of the arc fuses both the electrode and the work to form a weld.*

METALLIZING: *A method of overlay or metal bonding to repair worn parts.*

MIXING CHAMBER: *That part of a welding or cutting torch in which the gases are mixed for combustion.*

MULTI-IMPULSE WELDING: *The making of spot, projection, and upset welds by more than one impulse of current. When alternating current is used each impulse may consist of a fraction of a cycle or a number of cycles.*

NEUTRAL FLAME: *A gas flame in which the oxygen and acetylene volumes are balanced and both gases are completely burned.*

NICK BREAK TEST: *A method for testing the soundness of welds by nicking each end of the weld, then giving the test specimen a sharp hammer blow to break the weld from nick to nick. Visual inspection will show any weld defects.*

NONFERROUS: *Metals which contain no iron. Aluminum, brass, bronze, copper, lead, nickel, and titanium are nonferrous.*

NORMALIZING: *Heating iron-base alloys to approximately 100 °F (38 °C) above the critical temperature range followed by cooling to below that range in still air at ordinary temperature.*

NUGGET: *The fused metal zone of a resistance weld.*

OPEN CIRCUIT VOLTAGE: *The voltage between the terminals of the welding source when no current is flowing in the welding circuit.*

OVERHEAD POSITION: *The position in which welding is performed from* *the underside of a joint and the face of the weld is approximately horizontal.*

OVERLAP: *The protrusion of weld metal beyond the bond at the toe of the weld.*

OXIDIZING FLAME: *An oxyacetylene flame in which there is an excess of oxygen. The unburned excess tends to oxidize the weld metal.*

OXYACETYLENE CUTTING: *An oxygen cutting process in which the necessary cutting temperature is maintained by flames obtained from the combustion of acetylene with oxygen.*

OXYACETYLENE WELDING: *A welding process in which the required temperature is attained by flames obtained from the combustion of acetylene with oxygen. OXY-ARC CUTTING: An oxygen cutting process in which the necessary cutting temperature is maintained by means of an arc between an electrode and the base metal.*

OXY-CITY GAS CUTTING: *An oxygen cutting process in which the necessary cutting temperature is maintained by flames obtained from the combustion of city gas with oxygen.*

OXYGEN CUTTING: *A process of cutting ferrous metals by means of the chemical action of oxygen on elements in the base metal at* **elevated temperatures.**

OXYGEN GOUGING: *An application of oxygen cutting in which a chamfer or groove is formed.*

OXY-HYDROGEN CUTTING: *An oxygen cutting process in which the necessary cutting temperature is maintained by flames obtained from the combustion of city gas with oxygen.*

OXY-HYDROGEN WELDING: *A gas welding process in which the required welding temperature is attained by flames obtained from the combustion of hydrogen with oxygen.*

OXY-NATURAL GAS CUTTING: *An oxygen cutting process in which the necessary cutting temperature is maintained by flames obtained by the combustion of natural gas with oxygen.*

OXY-PROPANE CUTTING: *An oxygen cutting process in which the necessary cutting temperature is maintained by flames obtained from the combustion of propane with oxygen.*

PASS: *The weld metal deposited in one general progression along the axis of* *the weld.*

PEENING: *The mechanical working of metals by means of hammer blows. Peening tends to stretch the surface of the cold metal, thereby relieving contraction stresses.*

PENETRATE INSPECTION: *a. Fluorescent. A water washable penetrate with high fluorescence and low surface tension. It is drawn into small surface openings by capillary action. When exposed to black light, the dye will fluoresce. b. Dye. A process which involves the use of three non-corrosive liquids. First, the surface cleaner solution is used. Then the penetrate is applied and allowed to stand at least 5 minutes. After standing, the penetrate is removed with the leaner solution and the developer is applied. The dye penetrate, which has remained in the surface discontinuity, will be drawn to the surface by the developer resulting in bright red indications.*

PERCUSSIVE WELDING: *A resistance welding process in which a discharge of electrical energy and the application of high pressure occurs simultaneously, or with the electrical discharge occurring slightly before the application of pressure.*

PERLITE: *Perlite is the lamellar aggregate of ferrite and iron carbide resulting from the direct transformation of austenite at the lower critical point.*

PITCH: *Center to center spacing of welds.*

PLUG WELD: *A weld is made in a hole in one member of a lap joint, joining that member to that portion of the surface of the other member which is exposed through the hole. The walls of the hole may or may not be parallel, and the hole may be partially or completely filled with the weld metal.*

POKE WELDING: *A spot welding process in which pressure is applied manually to one electrode. The other electrode is clamped to any part of the metal much in the same manner that arc welding is grounded.*

POROSITY: *The presence of gas pockets or inclusions in welding.*

POSITIONS OF WELDING: *All welding is accomplished in one of four positions: flat, horizontal, overhead, and vertical. The limiting angles of the various positions depend somewhat as to whether the weld is a fillet or groove weld.*

POST-HEATING: *The application of heat to an assembly after a welding,* brazing, soldering, thermal spraying, or cutting operation.

POST-WELD INTERVAL: *In resistance welding, the heat time between the end of weld time, or weld interval, and the start of hold time. During this interval, the weld is subjected to mechanical and heat treatment.*

PREHEATING: *The application of heat to a base metal prior to a welding or cutting operation.*

PRESSURE CONTROLLED WELDING: *The making of a number of spot or projection welds in which several electrodes function progressively under the control of a pressure sequencing device.*

PRESSURE WELDING: *Any welding process or method in which pressure is used to complete the weld.*

PRE-WELD INTERVAL: *In spot, projection, and upset welding, the time between the end of squeeze time and the start of weld time or weld interval during which the material is preheated. In flash welding, it is the time during which the material is preheated.*

PROCEDURE QUALIFICATION: *The demonstration that welds made by a specific procedure can meet* **prescribed standards.**

PROJECTION WELDING: *process between two or more surfaces or between the ends of one member and the surface of another. The welds are localized at predetermined points or projections.*

PULSATION WELDING: *A spot, projection, or seam welding process in which the welding current is interrupted one or more times without the release of pressure or change of location of electrodes.*

PUSH WELDING: *The making of a spot or projection weld in which the force is aping current is interrupted one or more times without the release of pressure or change of location of electrodes.*

PUSH WELDING: *The making of a spot or projection weld in which the force is applied manually to one electrode and the work or a backing bar takes the place of the other electrode.*

QUENCHING: *The sudden cooling of heated metal with oil, water, or compressed air.*

A resistance welding **REACTION STRESS:** *The residual stress which could not otherwise exist if the members or parts being welded were isolated as free bodies without connection to other parts of the structure.*

REDUCING FLAME: *See CARBURIZING FLAME.*
REGULATOR: *A device used to reduce cylinder pressure to a suitable torch working pressure.*

REINFORCED WELD: *The weld metal built up above the surface of the two abutting sheets or plates in excess of that required for the size of the weld specified.*

RESIDUAL STRESS: *Stress remaining in a structure or member as a result of thermal and/or mechanical treatment.*

RESISTANCE BRAZING: *A brazing process in which bonding is produced by the heat obtained from resistance to the flow of electric current in a circuit of which the workpiece is a part, and by using a nonferrous filler metal having a melting point above 800 °F (427 °C), but below that of the base metals. The filler metal is distributed in the joint by capillary attraction.*

RESISTANCE BUTT WELDING: *A group of resistance welding processes in which the weld occurs simultaneously over the entire contact area of the parts being joined.*

RESISTANCE WELDING: *A group of welding processes in which fusion is produced by heat obtained from resistance to the flow of electric current in a circuit of which the workpiece is a part and by the application of pressure.*

REVERSE POLARITY: *The arrangement of direct current arc welding leads in which the work is the negative pole and the electrode is the positive pole of the welding arc.*

ROCKWELL HARDNESS TEST: *In this test a machine measures hardness by determining the depth of penetration of a penetrator into the specimen under certain arbitrary fixed conditions of test. The penetrator may be either a steel ball or a diamond spherocone.*

ROOT: *See ROOT OF JOINT and ROOT OF WELD.* **ROOT CRACK:** *A crack in the weld or base metal which occurs at the root of a weld.*

ROOT EDGE: *The edge of a part to be welded which is adjacent to the root.*
ROOT FACE: *The portion of the prepared edge of a member to be joined by a groove weld which is not beveled or grooved.*

ROOT OF JOINT: *That portion of a joint to be welded where the members approach closest to each other. In cross section, the root of a joint may be a point, a line, or an area.*

ROOT OF WELD: *The points, as shown in cross section, at which the bottom of the weld intersects the base metal surfaces.*

ROOT OPENING: *The separation between the members to be joined at the root of the joint.*

ROOT PENETRATION: *The depth a groove weld extends into the root of a joint measured on the centerline of the root cross section.*

SCARF: *The chamfered surface of a joint.*

SCARFING: *A process for removing defects and checks which develop in the rolling of steel billets by the use of a low velocity oxygen deseaming torch.*

SEAL WELD: *A weld used primarily to obtain tightness and to prevent leakage. SEAM WELDING: Welding a lengthwise seam in sheet metal either by abutting or overlapping joints.*

SELECTIVE BLOCK SEQUENCE: *A block sequence in which successive blocks are completed in a certain order selected to create a predetermined stress pattern.*

SERIES WELDING: *A resistance welding process in which two or more welds are made simultaneously by a single welding transformer with the total current passing through each weld.*

SHEET SEPARATION: *In spot, seam, and projection welding, the gap surrounding the weld between faying surfaces, after the joint has been welded.*

SHIELDED WELDING: *An arc welding process in which protection from the atmosphere is obtained through use of a flux, decomposition of the electrode covering, or an inert gas.*

SHOULDER: *See ROOT FACE.*
SHRINKAGE STRESS: *See RESIDUAL STRESS.*

SINGLE IMPULSE WELDING: *The making of spot, projection, and upset welds by a single impulse of current. When alternating current is used, an impulse may consist of a fraction of a cycle or a number of cycles.*

SIZE OF WELD: *a. Groove weld. The joint penetration (depth of chamfering plus the root penetration when specified). b. Equal leg fillet welds. The leg length of the largest isosceles right triangle which can be inscribed within the fillet weld cross section. c. Unequal leg fillet welds. The leg length of the largest right triangle which can be inscribed within the fillet* weld cross section. d. Flange weld. The weld metal thickness measured at the root of the weld.

SKIP SEQUENCE: *See WANDERING SEQUENCE.*

SLAG INCLUSION: *Non-metallic solid material entrapped in the weld metal or between the weld metal and the base metal.*

SLOT WELD: *A weld made in an elongated hole in one member of a lap or tee joint joining that member to that portion of the surface of the other member which is exposed through the hole. The hole may be open at one end and may be partially or completely filled with weld metal. (A fillet welded slot should not be construed as conforming to this definition.)*

SLUGGING: *Adding a separate piece or pieces of material in a joint before or during welding with a resultant welded joint that does not comply with design drawing or specification requirements.*

SOLDERING: *A group of welding processes which produce coalescence of materials by heating them to suitable temperature and by using a filler metal having a liquidus not exceeding 450 °C (842 °F) and below the solidus of the base materials. The filler metal is distributed between the closely fitted surfaces of the joint by capillary action.*

SOLIDUS: *The highest temperature at which a metal or alloy is completely solid.*

SPACER STRIP: *A metal strip or bar inserted in the root of a joint prepared for a groove weld to serve as a backing and to maintain the root opening during welding.*

SPALL: *Small chips or fragments which are sometimes given off by electrodes during the welding operation. This problem is especially common with heavy coated electrodes.*

SPATTER: *The metal particles expelled during arc and gas welding which do not form a part of the weld.*

SPOT WELDING: *A resistance welding process in which fusion is produced by the heat obtained from the resistance to the flow of electric current through the work pieces held together under pressure by electrodes. The size and shape of the individually formed welds are limited by the size and contour of the electrodes.*

SPRAY TRANSFER: *A type of metal transfer in which molten filler metal is* propelled axially across the arc in small droplets.

STAGGERED INTERMITTENT FILLET WELD: Two lines of intermittent welding on a joint, such as a tee joint, wherein the fillet increments in one line are staggered with respect to those in the other line.

STORED ENERGY WELDING: The making of a weld with electrical energy accumulated electrostatically, electromagnetically, or electrochemically at a relatively low rate and made available at the required welding rate.

STRAIGHT POLARITY: The arrangement of direct current arc welding leads in which the work is the positive pole and the electrode is the negative pole of the welding arc.

STRESS RELIEVING: A process of reducing internal residual stresses in a metal object by heating to a suitable temperature and holding for a proper time at that temperature. This treatment may he applied to relieve stresses induced by casting, quenching, normalizing, machining, cold working, or welding.

STRING BEAD WELDING: A method of metal arc welding on pieces 3/4 in. (19 mm) thick or heavier in which the weld metal is deposited in layers composed of strings of beads applied directly to the face of the bevel.

STUD WELDING: An arc welding process in which fusion is produced by heating with an electric arc drawn between a metal stud, or similar part, and the other workpiece, until the surfaces to be joined are properly heated. They are brought together under pressure.

SUBMERGED ARC WELDING: An arc welding process in which fusion is produced by heating with an electric arc or arcs between a bare metal electrode or electrodes and the work. The welding is shielded by a blanket of granular, fusible material on the work. Pressure is not used. Filler metal is obtained from the electrode, and sometimes from a supplementary welding rod.

SURFACING: The deposition of filler metal on a metal surface to obtain desired properties or dimensions.

TACK WELD: A weld made to hold parts of a weldment in proper alignment until the final welds are made.

TEE JOINT: A joint between two members located approximately at right angles to each other in the form of a T.

TEMPER COLORS: *The colors which appear on the surface of steel heated at low temperature in an oxidizing atmosphere.*

TEMPER TIME: *In resistance welding, that part of the post weld interval during which a current suitable for tempering or heat treatment flows. The current can be single or multiple impulse, with varying heat and cool intervals.*

TEMPERING: *Reheating hardened steel to some temperature below the lower critical temperature, followed by a desired rate of cooling. The object of tempering a steel that has been hardened by quenching is to release stresses set up, to restore some of its ductility, and to develop toughness through the regulation or readjustment of the embrittled structural constituents of the metal. The temperature conditions for tempering may be selected for a given composition of steel to obtain almost any desired combination of properties.*

TENSILE STRENGTH: *The maximum load per unit of original cross-sectional area sustained by a material during the tension test.*

TENSION TEST: *A test in which a specimen is broken by applying an increasing load to the two ends. During the test, the elastic properties and the ultimate tensile strength of the material are determined. After rupture, the broken specimen may be measured for elongation and reduction of area.*

THERMITE CRUCIBLE: *The vessel in which the thermite reaction takes place.*

THERMITE MIXTURE: *A mixture of metal oxide and finely divided aluminum with the addition of alloying metals as required.*

THERMITE MOLD: *A mold formed around the parts to be welded to receive the molten metal.*

THERMITE REACTION: *The chemical reaction between metal oxide and aluminum which produces superheated molten metal and aluminum oxide slag.*

THERMITE WELDING: *A group of welding processes in which fusion is produced by heating with superheated liquid metal and slag resulting from a chemical reaction between a metal oxide and aluminum, with or without the application of pressure. Filler metal, when used, is obtained from the liquid metal.*

THROAT DEPTH: *In a resistance welding machine, the distance from the centerline of the electrodes or platens to the nearest point of interference for flatwork or sheets. In a seam welding machine with a universal head, the throat depth is measured with the machine arranged for transverse welding.*

THROAT OF FILLET WELD: *a. Theoretical. The distance from the beginning of the root of the joint perpendicular to the hypotenuse of the largest right triangle that can be inscribed within the fillet-weld cross section. b. Actual. The distance from the root of the fillet weld to the center of its face.*

TOE CRACK: *A crack in the base metal occurring at the toe of the weld.*
TOE OF THE WELD: *The junction between the face of the weld and the base metal.*

TORCH: *See CUTTING TORCH or WELDING TORCH.*

TORCH BRAZING: *A brazing process in which bonding is produced by heating with a gas flame and by using a nonferrous filler metal having a melting point above 800 °F (427 °C), but below that of the base metal. The filler metal is distributed in the joint of capillary attraction.*

TRANSVERSE SEAM WELDING: *The making of a seam weld in a direction essentially at right angles to the throat depth of a seam welding machine.*

TUNGSTEN ELECTRODE: *A non-filler metal electrode used in arc welding or cutting, made principally of tungsten.*

UNDERBEAD CRACK: *A crack in the heat affected zone not extending to the surface of the base* **metal.**

UNDERCUT: *A groove melted into the base metal adjacent to the toe or root of a weld and left unfilled by weld metal.*
UNDERCUTTING: *An undesirable crater at the edge of the weld caused by poor weaving technique or excessive welding speed.*

UPSET: *A localized increase in volume in the region of a weld, resulting from the application of pressure.*

UPSET WELDING: *A resistance welding process in which fusion is produced simultaneously over the entire area of abutting surfaces, or progressively along a joint, by the heat obtained from resistance to the flow of electric current through the area of contact of those surfaces. Pressure is* *applied before heating is started and is maintained throughout the heating period.*

UPSETTING FORCE: *The force exerted at the welding surfaces in flash or upset welding. V*

VERTICAL POSITION: *The position of welding in which the axis of the weld is approximately vertical. In pipe welding, the pipe is in a vertical position and the welding is done in a horizontal position.*

WANDERING BLOCK SEQUENCE: *A block welding sequence in which successive weld blocks are completed at random after several starting blocks have been completed.*

WANDERING SEQUENCE: *A longitudinal sequence in which the weld bead increments are deposited at random.*

WAX PATTERN: *Wax molded around the parts to be welded by a thermite welding process to the form desired for the completed weld.*

WEAVE BEAD: *A type of weld bead made with transverse oscillation.*

WEAVING: *A technique of depositing weld metal in which the electrode is oscillated. It is usually accomplished by a semicircular motion of the arc to the right and left of the direction of welding. Weaving serves to increase the width of the deposit, decreases overlap, and assists in slag formation.*

WELD: *A localized fusion of metals produced by heating to suitable temperatures. Pressure and/or filler metal may or may not be used. The filler mal has a melting point approximately the same or below that of the base metals, but always above 800 °F (427 °C).*

WELD BEAD: *A weld deposit resulting from a pass.*

WELD GAUGE: *A device designed for checking the shape and size of welds.*

WELD METAL: *That portion of a weld that has been melted during welding.*

WELD SYMBOL: *A picture used to indicate the desired type of weld.*

WELD-ABILITY: *The capacity of a material to form a strong bond of adherence under pressure or when solidifying from a liquid.*

WELDER CERTIFICATION: *Certification in writing that a welder has produced welds meeting prescribed standards.*

WELDER PERFORMANCE QUALIFICATION: *The demonstration of a welder's ability to produce welds meeting prescribed standards.*

WELDING LEADS: a. Electrode lead. The electrical conductor between the source of the arc welding current and the electrode holder. b. Work lead. The electrical conductor between the source of the arc welding current and the workpiece.

WELDING PRESSURE: The pressure exerted during the welding operation on the parts being welded.

WELDING PROCEDURE: The detailed methods and practices including all joint welding procedures involved in the production of a weldment.

WELDING ROD: Filler metal in wire or rod form, used in gas welding and brazing processes and in those arc welding processes in which the electrode does not provide the filler metal.

WELDING SYMBOL: The assembled symbol consists of the following eight elements, or such of these as are necessary: reference line, arrow, basic weld symbols, dimension and other data, supplementary symbols, finish symbols, tail, specification, process, or other references.

WELDING TECHNIQUE: The details of a manual, machine, or semiautomatic welding operation which, within the limitations of the prescribed joint welding procedure, are controlled by the welder or welding operator.

WELDING TIP: The tip of a gas torch especially adapted to welding.

WELDING TORCH: A device used in gas welding and torch brazing for mixing and controlling the flow of gases.

WELDING TRANSFORMER: A device for providing current of the desired voltage.

WELDMENT: An assembly whose component parts are formed by welding.

WIRE FEED SPEED: The rate of speed in mn/sec or in./min at which a filler metal is consumed in arc welding or thermal spraying.

WORK LEAD: The electric conductor (cable) between the source of arc welding current and the workpiece.

X-RAY: A radiographic test method used to detect internal defects in a weld.

YIELD POINT: The yield point is the load per unit area value at which a marked increase in deformation of the specimen occurs with little or no increase of load; in other words, the yield point is the stress at which a marked increase in strain occurs with little or no increase in stress.

www.ingramcontent.com/pod-product-compliance
Lightning Source LLC
Chambersburg PA
CBHW030402160726
47992CB00007B/2926